JOHN P. GIESY

# ACID RAIN INFORMATION BOOK

ACID RAIN INFORMATION BOOK

by

Frank A. Record
David V. Bubenick
Robert J. Kindya

GCA Corporation
GCA/Technology Division
Bedford, Massachusetts

NOYES DATA CORPORATION
Park Ridge, New Jersey, U.S.A.
1982

Library of Congress Catalog Card Number: 81-22323
ISBN: 0-8155-0892-1
Printed in the United States

Published in the United States of America by
Noyes Data Corporation
Mill Road, Park Ridge, New Jersey 07656

Library of Congress Cataloging in Publication Data

Record, Frank.
  Acid rain information book.

  Bibliography: p.
  Includes index.
  1. Acid rain. I. Bubenick, David V. II. Kindya,
Robert J. III. Title.
TD196.A25R4   363.7'394   81-22323
ISBN 0-8155-0892-1   AACR2

Foreword

This book discusses the major aspects of the acid rain problem which exists today; it points out the areas of uncertainty and summarizes current and projected research by various government agencies and other concerned organizations.

Acid rain, caused by the emission of sulfur and nitrogen oxides to the atmosphere and their subsequent transformation to sulfates and nitrates, is one of the most widely publicized and emotional environmental issues of the day. The potential consequences of increasingly widespread acid rain demand that this not altogether surprising phenomenon be carefully evaluated. This review of over 350 literature sources reveals a rapidly growing body of knowledge, but also indicates major gaps in understanding which should be narrowed, and unanswered questions which should be resolved.

The book is organized by the logical progression from sources of acid rain precursors—such as SO_x and NO_x from fossil fuel combustion in industrial processing, electric power production, and the transportation sector; to atmospheric transport, possible chemistry of transformation, and deposition processes; to effects, both adverse and beneficial, on aquatic and terrestrial ecosystems, construction materials, and humans. Possible mitigative measures and further research needs are considered. Finally, a chapter is included on current and proposed research by the U.S. Department of Energy, the U.S. Environmental Protection Agency, and the Electric Power Research Institute.

The information in the book is from *Acid Rain Information Book* (DOE Report EP-0018), by Frank A. Record, David V. Bubenick, and Robert J. Kindya of GCA Corporation, GCA/Technology Division, Bedford MA, prepared for the U.S. Department of Energy, May 1981.

The table of contents is organized in such a way as to serve as a subject index and provides easy access to the information contained in the book.

> Advanced composition and production methods developed by Noyes Data are employed to bring this durably bound book to you in a minimum of time. In order to keep the price of this book to a reasonable level, it has been partially reproduced by photo-offset directly from the original report and the cost savings passed on to the reader. Due to this method of publishing, certain portions of a few of the figures may be less legible than desired.

Acknowledgments

This document is the result of the combined efforts of many staff members of the GCA/Technology Division. Responsibility for the report was divided as follows: Section 2—Sources—was prepared by David V. Bubenick, assisted by Douglas R. Roeck, John A. Dirgo, Liyang Chu, and Hans A. Klemm; Section 3—Atmospheric Transport, Chemical Transformation, and Deposition Processes—by Frank A. Record, Michael A. Wojcik, and Sue Ellen Haupt; Section 4—Adverse and Beneficial Effects of Acid Precipitation, and part of Section 5—Mitigative Strategies—by Robert J. Kindya; and Section 7—Current and Projected Research—by Lisa A. Baci. The portion of Section 5 entitled "Regulatory Alternatives Evaluated" was summarized by DOE staff members from an ICF, Inc. preliminary report on utility SO_2 control strategies. Section 6 was organized by David Bubenick with joint contributions from all of the project members. Project coordination was headed by Dr. Record, assisted by Mr. Bubenick.

The authors wish to thank Robert Kane, DOE Project Officer, for his assistance in assembling material for their review and for his continuing guidance throughout the project. The collection of reference material was also greatly expedited by GCA's librarian, Josephine Silvestro. The authors wish to acknowledge, with appreciation, the careful review of the first draft of this document provided by DOE staff members from the following offices: Environment, Policy and Evaluation, Fossil Energy, Economic Regulatory Administration, and General Council.

Notice

The material in this book was prepared as an account of work sponsored by the U.S. Department of Energy. Publication does not signify that the contents necessarily reflect the views and policies of the contracting agency or the publisher, nor does mention of trade names or commercial products constitute endorsement or recommendation for use.

Table of Contents

EXECUTIVE SUMMARY...1
  Introduction..1
    Acid Precipitation and Its Measurement 1
    Role of Atmospheric and Terrestrial Systems 2
    Organization...2
  Sources...2
    Natural Sources of SO_x and NO_x...2
    Anthropogenic Sources of SO_x and NO_x Emissions..........................4
    Other Sources...8
  Atmospheric Transport, Transformation, and Deposition Processes 9
    Atmospheric Chemistry and Physics..9
      Chemical Transformation During Transport 9
      Removal of Pollutants by Precipitation..10
    Regional Transport and Deposition Modeling 10
    International Aspects of Transport 11
  Trends in Acidity 11
    Long-Term Trends in Precipitation pH 11
    Cumulative Buildup of Acidic Components.......................................13
  Adverse and Beneficial Effects of Acid Precipitation...................................15
    Effect of Acidic Precipitation on Aquatic Ecosystems 16
      Acidification of Lakes..16
      Effects on Fish 16
      Effects on Plant Life and the Food Chain.....................................17
      Effects on Microorganisms and Decomposition.............................18
      Effects on Other Aquatic Organisms...18
    Effect of Acidic Precipitation on Terrestrial Ecosystems.......................18
      Effects on Vegetation..18
      Effects on Soil...19
    Effects of Acidic Precipitation on Materials 19
    Effects of Acidic Precipitation on Human Health 19
  Mitigative Strategies 20
  Summary of Issues, Concerns, and Further Research Needs 20
  Current and Proposed Research on Acid Precipitation 22
    Department of Energy..22
    Environmental Protection Agency...22
    Electric Power Research Institute 23

1. INTRODUCTION 24
  Acid Precipitation and Its Measurement 24
  Discovery of the Phenomenon...25
  Possible Effects...26
  Department of Energy and Environmental Protection Agency Interest and Involvement 26
    Department of Energy..27
    Environmental Protection Agency...28
  References..30

2. SOURCES 31
  Introduction 31

Natural Sources of SO_x and NO_x 32
  Natural Sources of SO_x 32
  Natural Sources of NO_x 33
Anthropogenic Sources of SO_x and NO_x Emissions 40
  Magnitude and Distribution 40
    Regional Summaries 47
    Historical Data 50
    Projections 50
    Local Sources 55
    Canadian Emission Inventory 57
  Factors Affecting Source Emissions 60
    Combustion Variables 60
    Control Technology 62
    Other Variables 62
  Quality of Data Base 63
Other Sources 67
  Other Sources Affecting Acid Rain Formation 67
    Ammonia 67
    Chlorides 70
  Synergistic Effects 72
    Ozone 72
    Carbon Dioxide 73
    Particulates 74
Summary 75
References 77

3. ATMOSPHERIC TRANSPORT, TRANSFORMATION, AND DEPOSITION PROCESSES 87
Meteorological Variability 87
  Seasonal Changes in Mean Flow 87
  Storm Tracks and Precipitation Patterns 88
  Occurrence of Stagnant Conditions 92
Atmospheric Chemistry and Physics 94
  Chemical Transformation during Transport 94
    Conversion of Nitrogen Oxides to Nitrates 96
  Removal of Pollutants by Precipitation 97
  Effects of Stack Height 98
  Relative Contributions of Wet and Dry Deposition 99
  Variability of Acidity Within and Among Storms 101
    Deposition as a Function of Precipitation Amount 102
    Spatial Variability Within a Metropolitan Network 102
    Temporal Variation Within a Storm 104
  Topographic Impacts 105
Review of Regional Transport and Deposition Modeling 106
  The PNL Regional Model 106
  The SRI EURMAP-1 Regional Model 109
  The ASTRAP Regional Model 113
  Summary of State-of-the-Art Regional Modeling 121
  The Areas of Uncertainties and Needed Improvement in Regional Transport and
    Deposition Modeling 121
International Aspects of Transport 122
  Prevailing Meteorology 122
  Transboundary Flux Estimates 122
  International Research 125
  Global Component of Acid Rain 125
Interregional Differences 127
  Measurement of Acidic Precipitation 127
    Background 127
    Long-Term Trends 127
    Short-Term or Seasonal Trends 134
    Current Monitoring 134
    Contributing Components to Acidity 136
    Monitoring of Lakes 136
  Sources of Acidic Components 137
    Difficulties in Assessing Sources 137

Nonquantitative Methods 137
Modeling Results... 140
Local Sources of Acidic Components 142
Cumulative Buildup of Acidic Components............................. 142
Acid Buildup in Lakes.. 142
Area Sensitivity to Acidic Precipitation............................. 144
Models to Determine Acceptable Loadings of Acidic Materials......... 145
Summary... 147
References.. 149

4. ADVERSE AND BENEFICIAL EFFECTS OF ACID PRECIPITATION 156
Introduction.. 156
Effect of Acidic Precipitation on Aquatic Ecosystems 157
Acidification of Lakes.. 157
Effects on Fish 161
Effects on Plant Life and the Food Chain............................ 165
Effects on Microorganisms and Decomposition......................... 166
Effects of Other Aquatic Organisms 166
Effect of Acidic Precipitation on Terrestrial Ecosystems............ 167
Effects on Vegetation 170
Effects on Soil... 173
Effect of Acidic Precipitation on Materials 175
Effects of Acidic Precipitation on Human Health 176
Summary... 177
References.. 179

5. MITIGATIVE STRATEGIES.. 188
Emission Reductions... 188
Regulatory Alternatives Evaluated 188
Liming.. 192
References.. 200

6. SUMMARY OF ISSUES, UNCERTAINTIES AND FURTHER RESEARCH NEEDS 202
References.. 213

7. CURRENT AND PROPOSED RESEARCH ON ACID PRECIPITATION 214
Department of Energy.. 214
Environmental Protection Agency..................................... 217
Effect of Acid Precipitation on Aquatic and Terrestrial Ecosystems.. 223
Effects of Acid Precipitation on Crops and Forests.................. 223
Characterization and Quantification of the Transfer, Fate, and Effects of SO_x, NO_x
and Acid Precipitation on Forest Ecosystems Representative of the Tennessee
Valley Region 223
Effects of Acid Rain on Terrestrial Ecosystems...................... 224
The Electric Power Research Institute............................... 225
References.. 228

Executive Summary

INTRODUCTION

Acid rain is one of the most widely publicized environmental issues of the day. The potential consequences of increasingly widespread acid rain demand that this phenomenon be carefully evaluated. Review of the literature shows a rapidly growing body of knowledge, but also reveals major gaps in understanding that need to be narrowed. This book discusses major aspects of the acid rain phenomenon, points out areas of uncertainty, and summarizes current and projected research by responsible government agencies and other concerned organizations.

Acid Precipitation and Its Measurement

The free acidity of a solution such as rain is determined by the concentration of hydrogen ions present. It is commonly expressed in terms of a pH scale where pH is defined as the negative logarithm of the hydrogen ion concentration. The pH scale extends from 0 to 14, with a value of 7 representing a neutral solution. Values less than 7 indicate acid solutions; values greater than 7 indicate basic solutions. Because the scale is logarithmic, each whole number increment represents a tenfold change in acidity, thus, pH 4 is 10 times as acidic as pH 5 and 100 times as acidic as pH 6.

Precipitation includes all forms of water that condense from the atmosphere and fall to the ground. Unpolluted precipitation is frequently assumed to have a pH of 5.65, the same value as distilled water in equilibrium with carbon dioxide under laboratory conditions. Hence, the term acid rain has come to mean rainfall with a pH of less than 5.6.

Contaminants in the atmosphere can shift the pH either way. Entrained soil particles of the West and Midwest tend to be basic and can increase the pH. In contrast, soil particles from the eastern United States are frequently acidic. The presence of such acid particles, or other aerosols formed from the interaction with gaseous sulfuric or nitric acid, could lower the pH.

Precipitation removes gases and particles from the atmosphere by two processes: (1) rainout, which is the incorporation of material into cloud drops that grow in size sufficiently to fall to the ground, and (2) washout, which occurs when material below the cloud is swept out by rain or snow as it falls. Together, these two processes account for wet deposition of acidic material on the earth's surface. Atmospheric pollutants are also removed from the atmosphere in the absence of precipitation by direct contact with the ground and vegetation and by gravitational settling. This process is called dry deposition. Effects of the two types of deposition on the environment are indistinguishable. Both types are usually implicitly included under the popular term acid rain.

Role of Atmospheric and Terrestrial Systems

Figure 1 depicts the role of atmospheric and terrestrial systems in transporting and transforming acid precipitation precursor pollutants from natural and manmade sources to a variety of receptors. The introduction of contaminants into the atmosphere is followed by transport, dispersal, transformation, and deposition on the earth's surface. Here, acidic substances may have either adverse or beneficial effects on vegetation, and may accelerate the erosion of stone or the corrosion of metals. After deposition, acidic products may be buffered by alkaline soils or carried by seepage and runoff into lakes, thus, potentially lowering the pH of the water and affecting the aquatic ecosystem.

Organization

The remainder of this summary and the body of the subsequent sections are organized by the logical progression from sources of acid rain precursors to effects and possible mitigative measures. This information is followed by a discussion of uncertainty in the understanding of the acid rain phenomenon and discription of current research and development.

SOURCES

Although manmade emissions of sulfur and nitrogen air pollutants are considered to be a major source of acid rain precursors, no quantitative cause-effect relationship between pollutant emissions and the measured acidity of precipitation has yet been determined between individual and regional sources or source categories and receptor areas situated some distance downwind. This situation results from the very complex nature of many chemical and physical processes that are involved in the transformation , transport, and deposition of the complex mixture of substances comprising acidic precipitation. Contributions of possible acid-rain precursors from natural sources are also important. It is apparent that more work must be done on tracing the release of pollutants, determining their conversion and transport rates, and measuring their deposition rates in order to evaluate the effects of modification of the various source contributions on the acid precipitation in receptor areas.

Natural Sources of SO_x and NO_x

To assess the magnitude and distribution of anthropogenic contributions of SO_x and NO_x from individual sources or from source areas, a knowledge of background levels is necessary. A common approach consists of performing a mass balance of the pollutants in a known air volume using approximations of the emission, atmospheric transformation and transport, scavenging, and removal of the pollutant. Budgets have been especially useful in sulfur and nitrogen compound systems by allowing natural emission processes to be estimated.

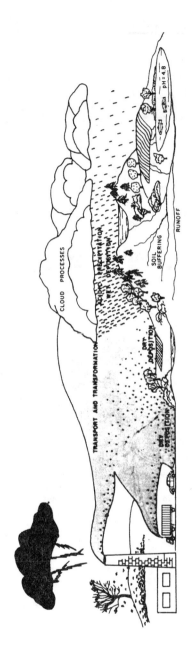

Figure 1. Schematic representation of the role of atmospheric and terrestrial systems in the acid rain phenomenon.

Although the magnitude of some natural sources of SO_x and NO_x may be significant, they are globally distributed, whereas manmade emissions tend to be much more concentrated.

The natural sources of SO_x are generally defined to include seaspray containing sulfates from oceans, organic compounds from bacterial decomposition of organic matter, reduction of sulfate in oxygen-depleted waters and soils, volcanoes, and forest fires. Although global and regional sulfur budgets have been estimated, few reliable measurements exist. The numbers for natural emissions have been lowered steadily from early estimates to about 38 million tons/year. The latest reported estimate attributes approximately two-thirds of all the sulfur emissions to man's activities. If the man-made contribution from the NEDS data base, which is discussed in depth below, is compared to natural source contributions, only 45% of total SO_x emissions are attributable to man's activities. In eastern North America, however, manmade emissions account for over 90 percent of the total SO_x emissions. Other studies have indicated similar levels for large areas of Europe.

Relative to global and regional SO_x emission estimates, the inventories for NO_x and related nitrogen compounds are much less certain, particularly for the fraction produced by natural sources. Global fluxes of nitrogen compounds are based largely on extrapolation of experimentally determined, small-scale emission factors or use of mass balances to obtain crude estimates for unknown sources. Estimated ratios of natural emissions of NO_x from terrestrial and aquatic sources to those from anthropogenic sources have ranged from 15:1 to 1:1 because of the variability in the natural source component estimate. Global estimates of natural NO_x emissions suggest that natural terrestrial sources of NO and NO_2 and tropospheric production of NO_x-N by lightning can be significant contributors to the total NO_x background. The principal source of gaseous NO_x in terrestrial systems appears to be chemical decomposition of nitrates.

Anthropogenic Sources of SO_x and NO_x Emissions

Based on the 1977 National Emissions Data Systems (NEDS), total emissions of SO_x and NO_x in the United States for 1977 are 31.5 and 21.7 million tons, respectively. Major source categories contributing to these two pollutants and their percentage of national emissions are given in Table 1. It can be seen that fossil fuel combustion (coal and oil) and industrial processes (principally primary metal production) are the largest contributing sources of SO_x emissions. In addition to fossil fuel (including coal, oil, and natural gas), transportation also contributes a significant portion of total U.S. NO_x emissions. The regional distribution of both of these pollutants is depicted in Table 2.

Although the magnitude of the SO_x and NO_x emission problem is important, of at least equal concern to the acid rain problem is how these emissions are distributed across the country. Emission density maps developed from the NEDS inventory show that most of the emissions occur in the eastern half of the country.

TABLE 1. CATEGORICAL DISTRIBUTION OF MANMADE U.S. SO_x AND NO_x EMISSIONS FOR 1977, IN MILLIONS OF TONS

Major source category	SO_x emissions (percent of U.S.)	NO_x emissions (percent of U.S.)
Stationary fuel combustion	24.9 (79.0)	11.0 (50.7)
- Coal	19.5 (61.9)	5.8 (26.7)
- Oil	4.8 (15.2)	2.2 (10.1)
- Gas	- -	2.7 (12.4)
Industrial processes	5.6 (17.8)	1.0 (4.6)
- Primary metals	2.8 (8.9)	0.06 (0.3)
- Petroleum	0.98 (3.1)	0.44 (2.0)
- Chemical manuf.	0.87 (2.8)	0.19 (0.9)
- Mineral products	0.70 (2.2)	0.23 (1.1)
Transportation	0.86 (2.7)	9.4 (43.3)
- Gasoline	0.23 (0.7)	6.0 (27.6)
- Diesel fuel	0.39 (1.2)	3.1 (14.3)
Total U.S.	31.5	21.7

TABLE 2. REGIONAL EMISSIONS OF SO_x AND NO_x COMPARED TO POPULATIONS (PERCENT OF U.S. TOTALS)

EPA region	States	Percent of U.S. population	Percent of U.S. SO_x total	Percent of U.S. NO_x total
I	CT, ME, MA, NH, RI, VT	5.6	2.1	3.4
II	NJ, NY, PR, VI	12.9	5.3	7.0
III	DE, DC, MD, PA, VA, WV	11.1	15.0	10.7
IV	AL, FL, GA, KY, MS, NC, SC, TN	16.3	21.5	17.5
V	IL, IN, MN, MI, OI, WI	20.6	29.0	23.0
VI	AR, LA, NM, OK, TX	10.3	9.0	17.0
VII	IA, KS, MO, NE	5.3	6.6	6.5
VIII	CO, MT, ND, SD, VT, WY	2.9	2.9	3.7
IX	AZ, CA, HI, NV, GU, AS	11.7	7.3	8.1
X	AK, ID, OR, WA	3.3	1.3	3.1

Emission trends for SO_x and NO_x are shown in Figure 2. Although historical SO_x and NO_x emissions data from 1970 to 1977 are more reliable than those for the years 1940, 1950, and 1960, both pollutants show a significant increase from 1960 to 1970. Sulfur dioxide reached its peak in 1970 with 32.1 million tons/yr, whereas NO_x rose to 25.1 million tons/yr in 1973. The large growth in utility emissions was compensated in part by reduction in industrial and residential/commercial emissions, produced by switches from coal to oil and gas in the 1940s and 1950s. Transportation, industrial, and utility sources have contributed most to increased NO_x emissions from 1940 to the present.

Based on future energy scenarios defined in the Second National Energy Plan (NEPII), it appears that transportation will continue to be a major source of NO_x even with the imposition of emission controls, with utility and industrial emissions continuing to increase or remain relatively constant for NO_x and SO_x respectively. For SO_x, variations in the NEP II scenarios are much larger because of the uncertainties of retrofit controls on existing power plants. These projections, of course, will depend greatly on the specific energy scenarios assumed for the next 20 years. Sizable increases (greater than 25 percent) in electric generating capacity are planned by 1989 for all but the three northeastern EPA Federal Regions (Regions I, II, and III). Historically, the Northeast's contribution to total U.S. SO_x and NO_x has been decreasing, and it appears that the extent of interregional transfer of airborne sulfur and nitrogen oxides produced by coal burning in regions lying to the west may be an important environmental issue for the northeast.

Because coal-fired boilers are major contributors of SO_x emissions, the sulfur content of this fuel is an important factor. Nearly all of the available sulfur is discharged in the form of SO_2 and SO_3 because of complete conversion and low sulfur retention in the coal ash. The formation of sulfates ($SO_4^=$) can occur under high excess oxygen levels, also an important factor in NO_x formation, which decreases with decreasing excess air. The type of boiler and the type of fuel are also important in the formation of both SO_x and NO_x. For example, higher SO_3 emissions result from oil-fired units than from coal-fired units, for a given amount of fuel sulfur. Emissions of significant quantities of sulfates is also common from fuel oil combustion, probably due to catalytic oxidation of trace metals in fuel oils.

The type and degree of application of SO_x and NO_x control technologies play an important role in pollution reduction. Most of the flue gas desulfurization (FGD) systems currently operating are applied to eastern high sulfur coal with efficiencies of up to 90 percent. Coal cleaning, dry sorbent FGD, and fluidized-bed combustion (FBC) technologies are also receiving considerable attention for reducing SO_x and in the case of FBC, also NO_x emissions.

Nitrogen oxide reduction technology generally consists more of changing operating techniques than application of hardware. These methods include: injection of cooled combustion products, steam, or water into the flame zone; reduction of the combustion air preheat temperature; extraction of heat from the flame zone; reduced furnace load; low excess air firing; staged combustion; and flue gas recirculation. Combinations of these techniques have resulted in NO_x reductions as much as 37, 48, and 60 percent for coal, oil, and gas-fired boilers, respectively, compared to older combustion designs.

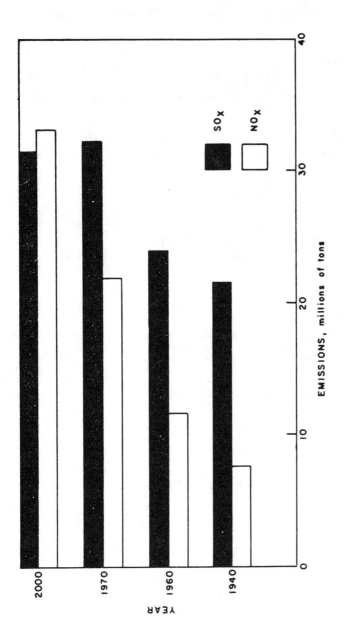

Figure 2. SO_x and NO_x emission trends.

Other Sources

In addition to the principal sources, "secondary" contributors or inhibitors are important in the formation of acid rain. Ammonia is generally found as an alkaline vapor able to neutralize either sulfuric or nitric acid in the atmosphere and will therefore tend to increase the pH of rain and snow. However, as ammonia dissolves to form ammonium (NH_4^+) the presence of this ion may increase the conversion rate of sulfur dioxide to sulfurous acid, and finally into sulfuric acid in the atmosphere. Most ammonia emissions are released into the atmosphere by natural and biological processes, such as the decay and decomposition of organic matter, forest fires, and volatilization from land and ocean masses. Anthropogenic sources account for a small percentage of the total ammonia emissions.

Average rainwater acidity in the northeastern United States has been calculated from representative samples to be 62 percent sulfuric acid, 32 percent nitric acid, and 6 percent hydrochloric acid (HCl). Despite its low but significant percentage in precipitation, HCl is a strong acid whose sources and mechanisms of formation have not been completely identified. The natural sources of chloride include salt spray from the oceans, volcanic gases, and upper atmospheric reactions. Anthropogenically produced chlorine and chlorides are emitted in various manufacturing and process operations; primarily in the manufacturing, handling, and liquefaction of chlorine gas and HCl. The combustion of coal by power generating facilities also releases chlorides into the atmosphere because central United States and Appalachian coals contain 0.01 to 0.5 percent chlorine by weight.

Ozone and other photochemical oxidants may play a role in the conversion of SO_x and NO_x to sulfates and nitrates, respectively. Atmospheric CO_2, through its influence on background pH, can determine the extent to which SO_x, NO_x, ammonia, and chlorides contribute to rain water acidification. Although the potential synergistic reactions of these substances are largely a matter of speculation and hypothesis at this time, further research can be expected to reveal the extent of their contributions to acid rain.

Another factor in the formation or neutralization of acid precipitation is the presence of natural and manmade dusts. Many natural dusts are alkaline and may react with and neutralize strong acids in the atmosphere. A similar role has been postulated for coal-fired boiler fly ash emissions, which are often alkaline. Some investigators have even suggested that fly ash removal by particulate control devices may have indirectly contributed to acid rain formation. Fly ash may play a more direct role in acid rain formation by catalytic oxidation of SO_2 by metallic constituents, such as vanadium pentoxide, or by absorption-oxidation in the presence of large amounts of water. Vanadium pentoxide is also formed in the combustion of residual oil and, therefore, may influence the fate of SO_2 from oil-fired power plants. The study of the catalytic oxidation of SO_2 to $SO_4^=$ needs much greater study, as it may be an important route in the transformation of SO_2 to $SO_4^=$ particulates. The conversion of SO_x and NO_x to their more stable particulate sulfate and nitrate forms increases their atmospheric lifetime, facilitates transport, and may contribute to the regional nature of the acid rain issue.

ATMOSPHERIC TRANSPORT, TRANSFORMATION, AND DEPOSITION PROCESSES

Manmade pollutants are injected into the atmosphere at heights ranging from a few feet, as is the case with auto exhaust, to more than 1000 feet in the case of tall stacks. Significant quantities of pollutants are also emitted from natural sources. The fate of all such pollutants depends on physical processes of dispersion, transport, and deposition and on complex chemical transformations that take place between source and final receptor. Their residence time within the atmosphere may be brief, as when emission takes place directly into an existing rainstorm, or may extend over several days or even weeks. In the latter case, the impact of pollutants, at least partially transformed, may extend to distances hundreds or even thousands of miles from the source. Methods to determine the relative contributions of possible sources to measured acidity are in early, exploratory stages. However, attempts are being made to describe the entire process using regional models.

Atmospheric Chemistry and Physics

Chemical Transformation During Transport--
Analyses of precipitation samples collected in North America indicate that their acidity is controlled to a great extent by the concentration of sulfate and nitrate ions. Although small amounts of sulfates and nitrates are emitted directly into the atmosphere, the major sources of these substances in precipitation are believed to be the oxidation end products of sulfur oxides and nitrogen oxides. The rate at which oxidation occurs between the point of emission and the receptor is therefore critical in determining the acidity of both wet and dry deposition products.

Conversion of sulfur dioxide to sulfate in the atmosphere may occur as a result of two types of reactions. In polluted atmospheres, homogeneous oxidation of sulfur dioxide proceeds after gas-phase collision with strong oxidizing radicals such as $HO\cdot$, $HO_2\cdot$, $CH_3O_2\cdot$. The source of these radicals is hydrocarbon-NO_x emissions, which through daytime photo-oxidation produces oxidizing radicals as intermediate products. The rate of oxidation is believed to depend on the initital ratio of hydrocarbons to NO_x, temperature, dewpoint, solar radiation, and the absolute concentrations of the reactive pollutants. Typical estimated rates range from 0.1 to 10 percent per hour.

The second type of reaction involves both gaseous and liquid or solid phases. Three heterogeneous mechanisms believed to be important in the atmospheric conversion of sulfur dioxide are: (1) catalytic oxidation in water droplets by transition metals; (2) oxidation in the liquid phase by strong photo-oxidation of hydrocarbon-NO_x mixtures; and (3) surface-catalyzed oxidation of sulfur dioxide on collision with solid particles, particularly elemental carbon (soot). Oxidation rates for these heterogeneous reactions in the atmosphere are not known.

The conversion of NO_x to nitric acid takes place through a series of complicated reactions during which nitrogen oxides switch back and forth between various stages of oxidation and eventually end up as nitrates. Because of the complexity of the chemical processes involved in the production of acidic products from nitrogen oxides in the atmosphere and the spatial and temporal variations of key parameters controlling these processes, rates of conversion of nitrogen oxides to nitrates can be expected to vary greatly. It has been estimated that these processes may take hours or days.

Removal of Pollutants by Precipitation--
Rainout processes begin with the condensation of water vapor on nuclei. Many of these nuclei are believed to be sulfate particles that have been formed as a result of the gas-to-aerosol conversion of sulfur dioxide emissions. Condensation is followed by droplet growth, during which various pollutants dissolve in the droplets, undergo chemical changes, and begin their descent to the ground in falling precipitation.

The washout of SO_2 by rain falling through a uniform concentration of SO_2 is a function of parameters such as the size spectrum of droplets and hence the rain rate, the initial pH of the rain, and the depth of the layer and the SO_2 concentration. Furthermore, the joint washout of SO_2 with other pollutants differs from that of SO_2 alone. In particular, the presence of NH_3 hastens the absorption of SO_2 and its conversion to sulfate.

Similar processes are involved in the scavenging of NO_x, nitrates, and other pollutants from the atmosphere.

Regional Transport and Deposition Modeling

Simulation of long-range transport and deposition with mathematical models can be a useful tool in analyzing the acid rain phenomenon. The Battelle, Pacific Northwest Laboratories PNL, Argonne National Laboratory ASTRAP, and the SRI International EURMAP-1 regional air pollution models have been applied to areas encompassing the northeastern United States and eastern Canada. All three models use objectively generated, regional-scale wind fields based on interpolation of available upper-air data to simulate transport and diffusion of atmospheric emissions. The PNL and EURMAP-1 models are based on a trajectory approach. The ASTRAP model, instead of simulating day-to-day transport events, uses a statistical approach to determine the location of long-term mean plumes. All three models assume that horizontal transport can be properly accounted for by a single wind field representative of the entire mixed layer.

At this time, because of a lack of understanding of the complex chemical processes taking place, particularly within clouds, and a limited knowledge of the wet removal processes, the models rely heavily on empirical relationships to describe transformation and removal processes. The greatest disagreement among model parameterizations exists in the wet removal mechanisms.

Preliminary model simulations have been limited to SO_2 and $SO_4^=$ transport and deposition. At the present, the nitrogen cycle has not been simulated, and models are in the testing and refining stage. A limited amount of comparison between model-calculated, monthly-averaged SO_2, $SO_4^=$, and total sulfur deposition to observation has been performed. The SO_2 and $SO_4^=$ concentration fields predicted by the models exhibited similar features. Their predicted concentration were generally within 30 percent of measured values in the vicinity of large point sources, however, predictions were off by a factor of 2 or more in rural parts of the modeled region. The PNL regional model wet deposition values agreed closely with a limited data base of four monitoring sites, being within 10 percent using the latest version of the model. The ASTRAP wet deposition results were compared to a more extensive data network. Predictions were often within 20 percent of observations, but the calculated distribution did not show the complex features visible in the data. As a result, predictions were at times off by more than a factor of 2. EURMAP-1 deposition predictions were not compared with data but the deposition fields were similar to the PNL and ASTRAP results.

International Aspects of Transport

In addition to simulating concentration fields and deposition patterns, the regional models can be used to estimate the United States-Canada transboundary flux.

Preliminary estimates by the ASTRAP model indicate that the United States contributes four to five times as much sulfur to Canada as it receives. As would be expected from the seasonal wind patterns, the net summer flux into Canada is greater than its winter counterpart. Also, a simple exponential and decay model estimates the difference between the United States-to-Canada and Canada-to-United States sulfur flux to be approximately a factor of 3. Seasonal transport estimates based on the EURMAP-1 model indicate that the ratio of United States-Canada transport to Canada-United States transport is 1.3 in the winter and 3.2 in the summer.

The ASTRAP and EURMAP 1 models estimate that Canadian source emissions contribute less than 5 percent of the total sulfur deposited in the northeastern United States. However, Canadian sources made a significant contribution in northern New York and northern New England. The model results also indicate that United States sources contribute about the same portion of total sulfur deposited in Eastern Canada as do Canadian sources.

TRENDS IN ACIDITY

Long-Term Trends in Precipitation pH

Because of the lack of long-term monitoring data, acidity trends in the United States have been poorly defined and necessarily based on calculated pH values. Cogbill and Likens pieced together precipitation data from the 1950s and 1960s and calculated the pH from ion concentrations. By comparing predicted pH values with actual measurements taken in the early 1970s, they estimated an error of 0.1 pH units. Maps illustrating their work appear in Figure 3. In the absence of more reliable data, their study has often been cited in estimating long-term trends. It shows both an increase in the area affected and lower pH precipitation appearing in New England and New York. However, the validity of these conclusions has been questioned by some investigators for the following reasons:

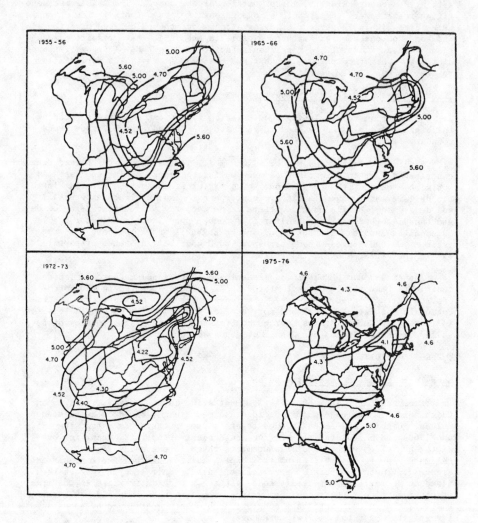

Figure 3. Isopleths of the weighted annual average pH of precipitation in the eastern United States in 1955-56, 1965-66, 1972-73, and 1976-76 (modified from Likens et al.).

- Sampling methods before 1972 did not associate acid precipitation with an event and collected total wet plus dry deposition.

- Precipitation samples were not stored correctly from the earlier time period and it has been noted that measurements taken in the field often differ significantly from laboratory measurements of the same sample.

- Calculated values have been assigned a margin of error of 0.5 pH units by one critic.

- The sampling sites for the different time periods were not the same.

Because pH has been found to vary between locations and storms, some critics have reanalyzed the data, examining trends at individual sites. The initial comparisons between 1955-56 and 1965-66 published by Cogbill and Likens had 10 common sites. Four of those showed increases in pH, two showed decreases, and the other four remained the same. Only two of the same sites were measured in the 1955-56/1972-73 comparison, one showing increasing pH, and the other showing a decrease. Between 1965-66 and 1972-73, eight stations were common, with pH increases at three, decreases at two, and no changes at three. It has been stated that if only the common stations were compared, no trend could be deduced.

The supposition of a trend toward increasing acidity has also been disputed on the grounds that the nine sites monitored by the United States Geological Survey from 1965 to 1978 in New York State where acid precipitation does occur have shown no significant trend over the time period. Time series of pH measurements at four of these sites are shown in Figure 4. Even the monthly data from Hubbard Brook, a site that has been referred to as recording increasing acidity over time, showed no strong evidence of trend. It has also been suggested that the apparent trend toward more acidic rain may be caused by factors other than the increase of strong acids in precipitation.

Cumulative Buildup of Acidic Components

The gradual accumulation of acidic components, either as a direct result of acid precipitation or from sources such as ground water, bays, polluted streams, mine drainage and fertilizer run off, can have an effect on water bodies. Most lakes have the ability to buffer or neutralize the incoming acids. In the softwater lakes that are sensitive to additions of strong acids, the predominant anion is bicarbonate (HCO_3^-), the dissociated form of the weak carbonic acid (H_2CO_3). As long as bicarbonates persist, the added acids are neutralized and the pH does not change. However, when the bicarbonate is exhausted, the pH quickly drops. This point is sometimes referred to as the threshold of lake acidification.

14   Acid Rain Information Book

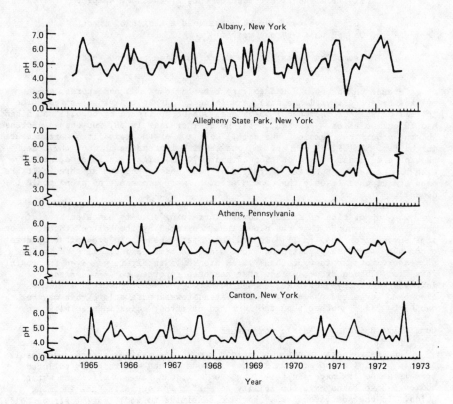

Figure 4. History of acidic precipitation at four USGS sites.

Not all lakes are equally susceptible to acidification. Lakes in New England, the Adirondacks, and northern Minnesota are showing signs of stress, whereas many of those sampled in western areas of the country that also receive acid precipitation have not experienced pH changes. Even in the same area, lakes display varying pH levels. The sensitivity of a given lake depends on many interacting factors in ways not yet well understood. The factors that are generally considered important are:

- meteorological patterns affecting precipitation,
- lake hydrology,
- watershed characteristics, and
- soil and bedrock geology.

ADVERSE AND BENEFICIAL EFFECTS OF ACID PRECIPITATION

This section discusses the potential impacts of acidification on the environment. Most of the data available to date on impacts of acidic precipitation are derived from studies of the effects of increased acidity on aquatic organisms. The effects of lowering pH on fish, plant species, and other members of fresh water ecosystems are relatively well documented; therefore, the manner and severity of disruption of the affected ecosystems produced by acidification may be postulated with some confidence.

Discussion of potential impacts of acidic precipitation on terrestrial ecology rests on more tenuous evidence. The terrestrial ecosystem is a complicated biological system, and deposition of acidic precipitation exerts a complex influence on the functioning of that ecosystem. The evaluation of potential impacts is complicated by the apparent trade-offs between benefits from nutrient enhancement and the possibility of inhibition of plant growth or other detrimental effects. Most of the effects data have been generated under laboratory or greenhouse conditions using simulations of exposure of terrestrial species to acidic precipitation.

In general, there has been no clear quantification of the magnitude of the potential adverse or beneficial impacts of acidic precipitation. Whether the observed adverse effects are a local or regional phenomenon caused by poor buffering capacity of the affected lakes or soils or whether the effects are more widespread is still unresolved. Past research has centered on those areas where effects, especially aquatic effects, have been observed. Such research has therefore involved the most acid-sensitive regions, systems, and organisms. Implied impacts of acid precipitation on more highly buffered areas whose acid resistance is higher are speculative at this point. It is possible that acidic precipitation is producing or will produce responses within the ecosystem even though it is not possible at the present time to observe, record, or evaluate them. Determining what these changes are, quantifying these changes, and determining whether the changes are harmful or beneficial can only be accomplished through systematic scientific investigation over a long period of time.

Effect of Acidic Precipitation on Aquatic Ecosystems

Acidification of Lakes--
The increasing acidity of freshwater lakes and streams appears to be a major environmental factor stressing aquatic ecosystems in Europe and North America. The chemical composition of lakes is largely determined by the composition of influents from precipitation and watershed drainage. Softwater lakes are usually produced by drainage of acidic igneous rocks, whereas hard waters contain large concentrations of alkaline earths derived from drainage of calcareous deposits. Hardness of water is associated with alkalinity and, therefore, with the increased capacity of the water to neutralize or buffer the acidity entering a lake. Chemical weathering and ion exchange are two mechanisms in watersheds that act to neutralize incoming acidity.

Depending on various factors, lakes exhibit a range of sensitivity to acidification. Included among these factors are the acidity of both wet and dry atmospheric deposition, the hydrology of the lake, the soil system, and the resultant chemistry of the surface water. The most important of these factors appear to be the soil system and associated canopy effects relative to the lake in question. Studies indicate that the capability of a lake and its drainage basin to neutralize the acidic inputs of precipitation is largely predicated by the composition of the bedrock of the watershed.

Although the mechanisms of acidic input into freshwater lakes and streams have been recognized, the magnitude of the contribution of acidic precipitation to lake acidification is far from resolved. Many studies have emphasized the complex nature of the interactions between precipitation and resultant water quality. Some authors caution that water quality effects usually attributed directly to the input of acidic precipitation could possibly be the result of lithospheric or ecosystem changes not caused by acid deposition. Some European investigators assign a secondary role to acidic precipitation in water quality changes. Others maintain that acidic precipitation is the causative factor.

Effects on Fish--
The death of fish in acidified freshwater lakes and streams has been more thoroughly studied, both in the laboratory and in the field, than any other aspect of lake and stream acidification. Various factors that affect the tolerance of fish to acidic waters have been identified, among which are species, strain, age, and size of the fish and physical factors including temperature, season, and hydrology. Species of fish vary in their tolerance to low pH. These data are based on experiments conducted with fish maintained at constant pH. Data are not available on species' response to transient pH changes.

The decline of fish populations in acidified lakes and streams has been reported in Scandinavia, Canada, and more recently in the United States. Although the disappearance of fish populations in Scandinavia was initially reported as long as 50 years ago, the rate of such disappearances has sharply increased during the past 15 years. In addition, investigators have documented the acidification of lakes and the loss of fish in Sudbury, Ontario and surveys conducted in New York State's Adirondack Mountains and in Pennsylvania have indicated an increase in the number of lakes and streams with acid pH (<5) over time.

Disappearance of fish from affected bodies of water is usually associated with one of two patterns. A sudden, short-term shift in pH resulting in acid shock may cause fish mortality. Sudden drops in pH could cause fish kills at pH levels above those normally toxic to fish. Such pH shocks often occur in early spring when snow melt releases acidic constituents accumulated during the winter.

A gradual decrease in pH with time is a second mechanism whereby acidification could result in elimination of fish populations. Field observations and laboratory experimentation have shown that prolonged acidity interferes with fish reproduction and spawning so that, over time, there is a decrease in fish population density and a shift in the size and age of the population to older and larger fish. This pattern has been observed in Norway, Sweden, Canada, and the United States.

Studies in the Adirondacks have indicated that mobilization of toxic metals, especially aluminum, is an additional factor that may contribute to mortality of fish at low pH levels. Soil leaching and mineral weathering by acidic precipitation may result in high concentrations of aluminum in surface and ground waters. Several studies in Sweden, Canada, and the United States, have revealed high mercury concentrations in fish from acidified regions. Elevated mercury levels in fish or fresh water lakes would have a potential detrimental impact on both aquatic species and human health. However, studies performed to date are far from comprehensive, and reports of results are still controversial.

Effects on Plant Life and the Food Chain--

Elimination or reduction in the fish population is the most obvious biological impact associated with acidification of freshwater lakes and streams. Less obvious, but of great importance, however, are the effects of acidification on other aquatic organisms. Organisms at all trophic levels within the food chain may be affected. Species can be reduced in number and variety, and primary production and decomposition may be impaired with a resultant disruption of the entire ecosystem.

Changes in pH have caused changes in the composition and structure of the aquatic plant communities involved in primary production. Lowering of pH of lakes in Ontario resulted in changes in species composition and in the standing crop and production of the phytoplankton community. The relative abundances of the algal flora also changed. In the lakes under study, differences in nutrient levels (phosphorus and nitrogen) were not responsible for these changes in primary productivity; acidity appeared to be the limiting factor.

Reductions in the diversity of the plant communities in lakes and streams and subsequent disruption in primary production reduces the supply of food and, therefore, the energy flow within the affected ecosystem. Changes in these communities also reduces the supply of minerals and nutrients. These factors limit the number of organisms that can exist within the ecosystem.

Effects on Microorganisms and Decomposition--
Acidification of lakes reduces microbiological activity, and, therefore, affects the rates of decomposition and the accumulation of organic matter in aquatic ecosystems. Organic matter in lakes plays a central role in the energetics of lake ecosystems. The biochemical transformations of detrital organic matter by microbial metabolism are fundamental to nutrient cycling and energy flux within the system, and the trophic relationships within lake ecosystems are almost entirely dependent on detrital structure.

Effects on Other Aquatic Organisms--
Invertebrate communities are also affected by acidification of freshwater lakes and streams. Surveys conducted at sites in Scandinavia and North America have shown that acidified lakes and streams have fewer species of bottom-dwelling invertebrates than do waters with higher pH.

Effect of Acidic Precipitation on Terrestrial Ecosystems

Assessing the impacts of acid precipitation on terrestrial ecosystems is extremely difficult. In fact, at present, it has not been possible to observe or measure changes in natural terrestrial ecosystems that could be directly attributed to acidic precipitation; however, such changes have been observed under controlled laboratory and field conditions. Therefore, it may be postulated that such effects could occur.

Effects on Vegetation--
Chemical species in the atmosphere reach plant surfaces through wet and dry deposition. Although sulfates, nitrates, and other water-soluble species may be assimilated through plant leaves, it has generally been assumed that the free hydrogen ion concentration in acidic precipitation is the component most likely to cause direct, harmful effects on vegetation. Experimental studies have supported this assumption, but there have been no reports of foliar symptoms on field-grown vegetation in the continental United States that could be attributed to exposure to ambient acidic precipitation.

A recent study, which is now partially completed, measured the effects of acid precipitation on 32 major crops that represent a total annual income in the United States of $50 billion. Crops were grown under controlled environmental conditions and exposure to simulated (sulfuric) acid rain of pH: 3.0, 3.5, and 4.0, in addition to a control rain of pH 5.7. Injury to foliage and effects on yield of edible portions were then determined. Preliminary results indicated that some crops suffered extensive damage, but others sustained little apparent injury, even under these severe exposure conditions.

Acidic precipitation can also cause indirect effects on plants and vegetation, some of them beneficial. Some studies have shown an increase in needle length and the weight of seedlings of Eastern white pine with increasing acidity of simulated precipitation. Investigators at the Argonne National Laboratory have reported no harmful effects on soybean productivity following exposure to simulated acidic rain. In fact, they observed a positive effect on productivity as reflected by seed growth.

Research to date on the effects of acidic precipitation on the leaching of chemical constituents from vegetation has resulted in contradictory results. It has been demonstrated that acidic precipitation can increase the leaching of various cations and organic carbon from the tree canopy. Foliar losses of nutrient cations from bean plants and maple seedlings were found to increase as acidity of the artificial mist to which they were exposed increased. However, in experiments using Norway spruce, researchers found no evidence of change in the foliar cation content although increased leaching was observed. It has been stated that increased leaching of nutrients from foliage can actually accelerate their uptake by plants. The impacts of the increased leaching of chemical substances from vegetation by acidic precipitation is still unresolved.

Effects on Soil--

Another area that suffers from limited investigation and inconclusive results is the effects or consequences of increased acidity on soil and the subterranean ecosystem. Effects have been postulated, but the picture is far from clear. It is especially difficult to factor out the impacts of acid precipitation on soil, if any, as compared to natural or anthropogenic mechanisms resulting in soil acidification, such as agricultural fertilization. Some authors contend that acid precipitation inputs to date are low compared to the possible influences of agricultural fertilization or liming practices.

Effects of Acidic Precipitation on Materials

Acidic precipitation can damage materials, structures, and manmade artifacts. It has the potential to accelerate corrosion of metals and erosion of stone. However, because a dominant factor in the formation of acidic precipitation can be sulfur compounds, it is difficult to distinguish effects of acidic precipitation from damage induced by sulfur pollution in general.

The influence of acidic precipitation on corrosion of metals has been investigated. Precipitation can have varying influence on corrosion. Rain may accelerate corrosion by forming a layer of moisture on the metallic surface and by adding hydrogen and sulfate ions. However, rain may also wash away sulfates deposited during dry deposition and can, therefore, retard corrosion. This problem has been investigated, and results indicated that mode of deposition complicates analysis of the impact of acid precipitation.

Acidic precipitation may leach chemical constituents from stonework just as acidic water leaches ions from soils and bedrock. However, at the present time, it is not possible to attribute observed effects of atmospheric sulfur compounds in general, or acidic precipitation in particular, to specific chemical compounds. The precise chemical mechanisms involved in such deteriorations are, likewise, unresolved. However, the effects are evident on buildings, monuments, and statuary.

Effects of Acidic Precipitation on Human Health

As previously observed, mobility of metallic compounds in soil is increased at low pH concentrations. Given this fact, there exists a potential indirect impact on human health through contamination of edible fish and drinking water supplies by these metallic species. However, comprehensive study and analysis of toxic metals in commercial or recreational fish catches has yet to be conducted. Also, reported concentration levels of these metallic species in analyzed waters have been orders of magnitude below public health drinking water standards.

MITIGATIVE STRATEGIES

One of the possible options for use in a program for the management of acid deposition may be to mitigate its harmful effects in susceptible areas. Methods suggested have included increasing the pH of affected lakes, soils, forests, etc.; the development of protective coatings for exposed structures and materials; and the development of acid-resistant species of crops, trees, and fish. Only the first approach, which involves liming of lakes and/or streams, has received any investigation thus far.

Liming has been attempted largely in Scandinavia and on a limited scale elsewhere. Although to date no observations of long-term detrimental effects of liming have been observed, the true ecological consequences are unknown. Invariably, alkalinity and pH will increase. Phosphorus release from lake sediments should be observed, probably as a result of ion-exchange processes with bicarbonate generated from liming. Concentrations of zinc, manganese, and aluminum drop with elevations in pH. Among the biological changes that have been reported subsequent to liming are increased phytoplankton and zooplankton diversity. In fish populations whose age distribution was skewed toward older groups as a result of acidification, liming resulted in restoration of younger age groups.

Treatment of affected waters by large-scale liming programs would represent major undertakings that could be logistically difficult and expensive. The economic realities involved in such programs in which the financial cost of the liming treatment is carefully balanced against the loss incurred if the treatment were not given must be evaluated in detail. This balance could be a major consideration in remote or generally inaccessible areas.

SUMMARY OF ISSUES, CONCERNS, AND FURTHER RESEARCH NEEDS

This section of the book summarizes the key issues, concerns, and further research needs relating to what is known of and speculated about acid rain. Table 3 is a condensed summary of the principal areas of uncertainty associated with the acid rain phenomenon. Table 6-1, at the end of Section 6 in the body of the book, is a rather detailed summary of the individual issues that have surfaced from a comprehensive review of the literature. These issues, which are discussed in detail throughout the book, are organized according to major subject areas, covering sources, atmospheric chemistry and physics, monitoring effects, and mitigative strategies. Corresponding to each issue is an indication of the level of concern as suggested in or directly inferred from the literature due to a lack of consensus among the experts. Examination of the issues has often led to identification of specific research needs and level of intensity that may be required to alleviate the underlying concern or uncertainty. The total resources required for these efforts are difficult to estimate because

TABLE 3. PRINCIPAL AREAS OF UNCERTAINTY ASSOCIATED WITH THE ACID RAIN PHENOMENON

Category	Extent of Information Gap	Will Research Help?	How Long Will Research Take
Emissions			
Man-Made			
o Global	Major	Yes	Years
o National	Minor	No	–
o Regional	Minor	No	–
o Local	Minor	No	–
Natural	Major	Yes	Years
Monitoring and Measurement			
o Evidence of Increased Acidity over Increasing Area	Major	Yes	Years
o Changes in Lake, Soil, Vegetation Acidity	Major	Yes	Years
o Susceptibility to Change	Major	Yes	Months
Regional Modeling			
o Neglect of NO_x Transformations	Major	Yes	Years
o Wet vs. Dry Deposition	Minor	Yes	Years
Effects			
o Fish Destruction	Minor	Yes	Months
o Lake Ecology	Major	Yes	Years
o Forest Damage	Major	Yes	Years
o Human Health	Minor	Yes	Months
o Research Time*	Major	Yes	Years

* Time available before major impacts are unavoidable.

of the complexity and multidisciplinary talents needed to comprehensively address each issue. One of the most important of these factors is that the character of anthropogenic emissions will be constantly changing as future energy scenarios, pollution abatement procedures, mandated control requirements, and industrial processes are implemented.

CURRENT AND PROPOSED RESEARCH ON ACID PRECIPITATION

Department of Energy

The DOE's acid rain research focuses on four areas: monitoring, atmospheric processes, ecological effects, and mitigation strategies. The DOE estimates that it will allocate over $860,000 for acid precipitation research in FY 1980 and over $1.3 million in FY 1981. Over one-third of the funding in both years is devoted to monitoring work, whereas less than a fourth of the funding goes to ecological effects research.

Currently, DOE is participating in several large acid precipitation monitoring projects. One of these, the MAP3S program, was initiated by ERDA (now DOE) as a major sulfate pollution study. The original purpose of the study was to simulate the atmospheric effects of emissions from fossil-fueled power plants. Another large monitoring program, being conducted by the Environmental Measurements Laboratory, focuses on the effects of changes in fuel-use patterns on the chemical composition of wet and dry deposition. The Department of Energy is also sponsoring several projects that examine the formation and effects of acid precipitation. Table 7-2 outlines several indirect, or supporting, acid rain research projects for FY 1980 and FY 1981.

In the future, DOE expects to expand its program in several areas including:

- precipitation chemistry data analysis;

- diagnostic and predictive source-receptor analyses and modeling using regional/national emissions and precipitation data bases;

- atmospheric chemistry and physics of acid rain formation;

- acid precipitation effects on plants and soils with emphasis on trace element nutrient availability and losses in selected terrestrial, aquatic, and estuarine ecosystems;

- synergistic effects of dry and wet deposited acid species on vegetation and mechanisms of their interaction; and

- assessment of mitigative strategies.

Environmental Protection Agency

EPA's current acid rain research focuses on essentially the same areas as DOE's: environmental effects; monitoring, and atmospheric processes. According to EPA's recently compiled inventory of Acid Rain Monitoring and Research

Projects, just under $4 million was devoted to acid rain research in FY 1979; funding in FY 1980 is expected to exceed $5 million. In contrast to DOE, EPA spent over half of its budget for acid rain research in FY 1979 and FY 1980 on environmental effects and economics research and devoted only 10 to 15 percent of its funding to monitoring projects.

EPA's research on environmental effects and economics examines a wide range of topics. Projects currently underway are investigating the impact of acid rain on agriculture, forests, and other terrestrial ecosystems, aquatic ecosystems, drinking and ground water supplies and materials' deterioration. In FY 1980, the agency is spending approximately $2.7 million to analyze the environmental consequences of acid precipitation. Table 7-3 outlines EPA's acid rain research program.

Electric Power Research Institute

The Electric Power Research Institute (EPRI) has spent over $5 million on acid rain research and anticipates spending another $10 to $15 million over the next 5 years. Several of EPRI's research projects have attracted considerable attention, particularly the Integrated Lake-Watershed Acidification Study and the study on the Fate of Atmospheric Emission Plume Trajectory over the North Sea.

EPRI estimates that it will spend over $13 million between 1980 and 1985 on ecological research alone. Most of this money is divided fairly evenly between studies in three areas: lake acidification, crop yield and quality, and forest yield and quality. EPRI also estimates it will spend between $1.5 and $1.6 million for studies in two other areas of ecological effects research: grassland yield and quality and aquatic biota. In addition to these projects, EPRI hopes to fund studies in the following areas:

- evaluation of nitrogen and SO_2 deposition on forests,
- effects of acid rain on aquatic biota, and
- effects of acid deposition on agricultural soils.

Section 1

Introduction

Acid rain is one of the most widely publicized environmental issues of the day. The potential consequences of increasingly widespread acid rain demand that the phenomenon be carefully evaluated. Review of the literature shows a rapidly growing body of knowledge, but also reveals major gaps in understanding that need to be narrowed. This book discusses major aspects of the acid rain phenomenon, points out areas of uncertainty, and summarizes current and projected research by responsible government agencies and other concerned organizations.

ACID PRECIPITATION AND ITS MEASUREMENT

The free acidity of a solution such as rain is determined by the concentration of hydrogen ions (H^+) present. It is commonly expressed in terms of a pH scale where pH is defined as the negative logarithm of the hydrogen ion concentration. The pH scale extends from 0 to 14, with a value of 7 representing a neutral solution. Values less than 7 indicate acid solutions; values greater than 7 indicate basic solutions. Because the scale is logarithmic, each whole number increment represents a tenfold change in acidity, thus, pH 4 is 10 times as acidic as pH 5 and 100 times as acidic as pH 6.

Natural precipitation includes all forms of water that condense from the atmosphere and fall to the ground. Unpolluted natural precipitation is frequently assumed to be slightly acidic with a pH of 5.65. This is the pH of distilled water in equilibrium with carbon dioxide, as determined under laboratory conditions. Whether 5.65 is the pH of unpolluted precipitation in nature has never been established. Nevertheless, many researchers today have accepted this assumption and refer to precipitation having a pH of less than 5.65 as acidic. Hence, the term acid rain has come to mean rainfall with a pH of less than 5.65.

Contaminants in the atmosphere can shift the pH either way. Soil particles in the West and Midwest frequently contain carbonates, tend to be basic, and can increase the pH. In contrast, entrained soil particles from the eastern U.S. are usually acidic. The presence of such acid particles, or other aerosols formed from interaction with gaseous sulfuric or nitric acid, would lower the pH. The processes affecting the acidity of precipitation are very numerous, and very complex. They include gas-to-particle transformations, photochemistry and catalytic chemistry, aqueous chemistry within cloud drops and precipitation, and regional and global atmospheric transport. The pH of precipitation is an integrated measure of the relative contributions of all of these complicated processes.

Precipitation removes gases and particles from the atmosphere by two processes: (1) rainout, which is the incorporation of material into cloud drops that grow in size sufficiently to fall to the ground, and (2) washout, which occurs when material below the cloud is swept out by rain or snow as it falls. Together, these two processes account for wet deposition of material on the earth's surface. Pollutants are also removed from the atmosphere in the absence of precipitation by direct contact with the ground and vegetation and by gravitational settling. This process is called dry deposition. Effects of the two types of deposition on the environment are indistinguishable. It should be pointed out, that with the exception of a few surfaces, even the measurement of dry deposition is very difficult.

DISCOVERY OF THE PHENOMENON

It has been known for many years that the chemical content of precipitation can vary as a result of the scavenging of various atmospheric gases and aerosols. Identification of the acid rain phenomenon and an understanding of its magnitude and widespread nature, however, awaited the establishment of an organized monitoring program in northern and western Europe during the 1950s.

An analysis of data collected by the European Air Chemistry Network that was carried out in 1966 showed an area of precipitation with values below pH 4.0 centered on the Low Countries. It also showed that the area of acid precipitation had expanded during the data collection period.[1] An overview of the changing chemistry of precipitation and surface waters in Europe during this period has been presented by Oden.[2]

The probability that a regional acid rain problem similar to that found in Europe also exists in the northeastern United States and adjacent parts of Canada was brought to the attention of the scientific community by Likens and Bormann in 1974.[3] Their conclusions were based primarily on 11-year records of precipitation chemistry in north-central New Hampshire (Hubbard Brook Ecosystem Study), 1970-1971 data at several New York sites, and scattered observations elsewhere. A more recent article by Likens et al.[4] includes isopleths that show a spread of acid rain in eastern North America between 1955-1956 and 1975-1976. Although the observations made in the United States clearly show that acid rain is falling, the frequently reported trend toward increasing acidity is largely inferred from composite data bases acquired by different sampling networks operated over different time periods and sometimes with different sampling methods. Some investigators who have analyzed these data have concluded that they are therefore inadequate to define trends in acidity during this period; for example, testimony recently presented before the Senate Committee on Environment and Public Works.[5] Resolution of the issue will require long-term operation of a high-density monitoring network and the use of uniform sampling and analytical procedures.

In addition to the observations that have been made in the eastern United States, more limited observations have been made in other parts of the United States and Canada. These data have confirmed the widespread occurrence of acid precipitation in both countries. Recently, measurements made in areas far from manmade sources of precursor pollutants (e.g., the windward side of the island of Hawaii) have shown high levels of acidity.[6] These results suggest the existence of a worldwide background level of acidity considerably greater than hitherto suspected.

POSSIBLE EFFECTS

When acid precipitation has been deposited on the ground, it may alter the composition of the soils and surface waters and exert a detrimental effect on indigenous plants and animals within the ecosystem. In addition, the deterioration of buildings and other corrosion effects could be accelerated. Understanding of the changes brought about by acid deposition is far from complete, but as increasing amounts of data are evaluated, there is general agreement that the effects are, on balance, detrimental to the environment. A commonly cited illustration of a change associated with the acidification of fresh waters is the decline and disappearance of fish populations (Wright et al.).[7] More subtle suspected effects include damage to other components of aquatic ecosystems, acidification and demineralization of soils, and reductions in crop and forest productivity. These effects can be cumulative or can result from peak acidity episodes (Glass et al.).[8] The extent of environmental damage is strongly dependent on the natural buffering capacity of the local soil. With regard to materials damage, it should be recognized that these effects are also produced by other chemical compounds and meteorological processes, as well as acid deposition.

DEPARTMENT OF ENERGY AND ENVIRONMENTAL PROTECTION AGENCY
INTEREST AND INVOLVEMENT

Over the past several years, the Federal government has become increasingly aware of the potential long-term threat to human health and the environment represented by acidic precipitation. Acid rain was described by the U.S. Department of Health, Education, and Welfare as one of the two most serious global environmental problems associated with fossil-fuel combustion, the other being the accumulation of carbon dioxide in the atmosphere.[9] As part of the National Energy Plan (NEP) in 1977, the President commissioned a study on potential environmental impacts of increased coal use. That report, known as the Rall Report,[10] identified acid rain in the United States as one of the six environmental problems requiring closer scrutiny. At present, insufficient knowledge exists on the explicit causes and total effects of acid rain, a fact that has hampered the identification and development of appropriate control and mitigation measures.

Several agencies within the Federal government have planned or have already initiated research efforts related to the acid rain phenomenon. These include the Departments of Agriculture (USDA), Interior, Energy, Commerce, the Environmental Protection Agency (EPA), the National Science Foundation (NSF), and the Council on Environmental Quality (CEQ). Several states have undertaken research efforts as well as industry, in particular, the Electric Power Research Institute (EPRI).

As a result of the President's August 2, 1979, Environmental Message directing a concerted national effort to understand the causes, magnitude, and impacts of acid rain and to identify measures that can mitigate acid rain impacts, a new initiative was proposed. The initiative called for the development of a Federal Acid Rain Assessment Plan. In addition to the Federal Agencies listed above, the membership of the coordination committee includes the State Department, the Office of Science and Technology Policy (OSTP), and the Tennessee Valley Authority (TVA). The committee is co-chaired by USDA and EPA, with CEQ coordinating the effort among the agencies. A revised version of the Federal Acid Rain Assessment Plan has been prepared (August 1980) and is currently being reviewed by committee members. The Acid Rain Coordination Committee is currently in the process of developing an inventory of Federally supported acid rain research in support of the Federal plan. In addition, two Acid Precipitation bills were introduced in Congress in 1979, S. 1754 and H.R. 5764. The Senate Bill became P.L. 96-294, the Energy Security Act, Title X. The purpose of this title is to provide for the identification of the causes and sources of acid precipitation; the evaluation of environmental, social, and economic effects of acid precipitation; and, based on these research results and within the guidelines of existing law and available control techniques, limit the identified emissions as well as mitigate the adverse effects which may result from acid precipitation. An interagency task force is being assembled to carry out this 10-year, $50 million study.

Department of Energy

As part of its responsibilities under the National Environmental Policy Act (NEPA), the Department of Energy (DOE) is required to assess options for generating power that result in minimal damage to human health and the national environment. Because electric power production is a major part of the U.S. energy economy and coal is the most abundant resource of domestic fossil energy, DOE is focusing on the implications of increased coal burning. The effects of emissions from coal-fired power plants are being investigated along with their relationship to emissions from all other sources with which they interact as well as their eventual transport and transformation into acidic components in acid precipitation. Some of the projects in the DOE acid rain research program are:

- MAP3S Precipitation Chemistry Network (although a major part of this work is now being conducted by EPA).
- Maintenance of a Rural Precipitation Chemistry Station at Whiteface Mountain.
- Atmospheric Pollution Scavenging.
- The Chemical Composition of Precipitation and Dry Deposition in the United States.
- Effects of Acid Precipitation on Forest Soils.

Recognizing that many of the environmental and health impacts of energy development are regional, the Energy Research and Development Administration (ERDA), forerunner of DOE, initiated a regional sulfate pollution study in 1976. This study, known as the Multistate Atmospheric Power Production Pollution Study[11] (MAP3S), had as its major goals the improvement of the Nation's capability to simulate the atmospheric effects of emissions from fossil-fuel electric generating plants. An eight-station MAP3S Precipitation Chemistry Network was established to (1) document the temporal and spatial extent of acid precipitation in the northeastern United States, (2) assist in developing a better understanding of acid rain formation mechanisms, (3) develop improved parameterizations of the wet removal process to be included in numerical simulation models, and (4) provide input to biological effects research projects.

Concurrent with the beginning of the investigations of the physics and chemistry of acid rain, several studies on the ecological effects of acid rain on plants and vegetation, soils, and water were started. In general, these studies were designed to provide information on the ability of representative ecosystems and plant species to cope with this pollutant loading and to begin to determine the true ecological and economic costs. These effects studies emphasized (1) the determination of the susceptibility of plants and vegetation to damage from simulated acid rain in laboratory studies, (2) the study of the role of acid rain on the cycling of trace elements and sulfur in the Walker Branch Watershed, a forested ecosystem, at Oak Ridge, Tennessee, and (3) the study of the effect of acid rain on New Hampshire forest soils. In addition, several projects addressed the interaction of dry deposited sulfur oxides and associated pollutants on soils, vegetation, and plant productivity.

In 1978 the Office of Management and Budget directed that DOE transfer $14 million of its health and environmental research to the EPA. This was accomplished at the beginning of Fiscal Year (FY) 1979. Most of the MAP3S program and the acid rain research being conducted by the national laboratories were included in that transfer. DOE will continue to co-fund the MAP3S Precipitation Chemistry Network, although at a reduced funding level.

Environmental Protection Agency

The EPA's program for investigating the acid deposition problem and constructing a data base for possible future regulatory action includes efforts in the areas of environmental effects, monitoring, and atmospheric processes.[12] Under the direction of EPA's Office of Research and Development (ORD), the program is being conducted in-house and through grants, interagency agreements, and contracts with universities and other institutions. Much of the data that has been developed has been incorporated into air pollutant criteria documents prepared by the Environmental Criteria and Assessment Office in Research Triangle Park, North Carolina. These documents are currently under intensive review by both government agencies and private industry.

The largest area of effort in acid rain research being pursued by EPA is environmental effects and economics. Programs are quite diversified, ranging from the evaluation of the effects of acid precipitation on aquatic and terrestrial ecosystems, crops and forests, soil bacteriological processes, and fish resources, to the assessment of economic benefits of acid rain control. The approaches taken vary from literature review assessments to experimental laboratory and field studies. Programs investigating the relationship between coal-fired power plant emissions and acid rain effects cover such topics as transport and transformation of stack emissions, deposition dynamics of plumes, and establishing the thresholds of ecological and physiological damage from single and combined doses of SO_x, ozone, and NO_x as precursors of acid rain.

One of the major monitoring and quality assurance programs being conducted through ORD is an anticipatory research program designed to provide an integrated and centralized data bank for all monitoring projects related to acid rain deposition. Results of the study will be used to identify major variables and to develop a reliable acid rain monitoring system. Other studies include development of a precipitation chemistry network to assess the impact of fossil-fuel emissions in the Northeast, an assessment of potential national and international acid rain impacts, and research into the mechanisms and major variables involved in dry deposition to natural surfaces.

As discussed earlier, the major EPA-sponsored project under the atmospheric processes program area is the MAP3S. The basic objective of this program is to define the relationships between the sources and characteristics of air pollutants, their deposition, and the chemical quality of precipitation. Another component of the program includes computer simulation modeling of the interaction between sources, transport mechanisms, and receptors of acid rain, emphasizing the role of sulfates in the Northeast. Field data are also being collected to guide the development and testing of regional-scale air quality models. Another major monitoring program is the National Atmospheric Deposition Plan (NADP), which is defined in detail in Chapter 7.

REFERENCES

1. Barnes, R. A. The Long Range Transport of Air Pollution - A Review of European Experience. JAPCA, 29(12):1219-1235, 1979.

2. Oden, S. The Acidity Problem - An Outline of Concepts. In: Proceedings of the First International Symposium on Acid Precipitation and the Forest Ecosystem. USDA Forest Service and Ohio State University, Columbus, Ohio, 1975.

3. Likens, G. E., and F. H. Bormann. Acid Rain: A Serious Regional Environmental Problem. Science, 184:1176-1179, 1974.

4. Likens, G. E., R. F. Wright, J. N. Galloway, and T. J. Butler. Acid Rain. Sci. Am., 241(4):43-51, 1979.

5. Perhac, R. M. Testimony for the Electric Power Research Institute Before the Subcommittee on Environmental Pollution of the Senate Committee on Environment and Public Works, 1980.

6. Mondaca, B. G., ed. Geophysical Monitoring for Climatic Change No. 7. Summary Report 1978. U.S. Department of Commerce, NOAA, Env. Res. Lab., Boulder, Colorado, 1979.

7. Wright, R. F., T. Dale, E. T. Gjessing, G. R. Hendrey, A. Henriksen, M. Johannessen, and I. P. Muniz. Impact of Acid Precipitation on Freshwater Ecosystems in Norway. In: Proceedings of the First International Symposium on Acid Precipitation and the Forest Ecosystem, USDA Forest Service and Ohio State University, Columbus, Ohio, 1975.

8. Glass, N. R., G. E. Glass, and P. J. Rennie. Effects of Acid Precipitation. Environ. Sci. Technology, 13:1350-1355, 1979.

9. U.S. Department of Health, Education, and Welfare. Report of the Committee on Health and Environmental Effects of Increased Coal Utilization. 1978.

10. U.S. Department of Health, Education, and Welfare. Report of the Committee on Health and Environmental Effects of Increased Coal Utilization. Washington, D.C., 1977.

11. MacCracken, M. C. The Multistate Atmospheric Power Production Pollution Study - MAP3S: Progress Report for FY-1977 and FY-1978. DOE/EV-0040, Department of Energy, July 1979.

12. EPA Completes Acid-Rain Research Inventory. Inside EPA, August 29, 1980. p. 12.

Section 2

Sources

INTRODUCTION

Although man-made emissions of sulfur and nitrogen air pollutants are considered to be major sources of acid rain precursors, no quantitative cause-effect relationship between pollutant emissions and the measured acidity of precipitation has yet been determined. Also, no direct cause-effect relationships have yet been determined between individual and regional sources or source categories and receptor areas situated some distance downwind from these sources. This situation results from the very complex nature of the many chemical and physical processes that are involved in the transformation, transport, and deposition of the complex mixture of substances comprising acidic precipitation. Contributions of possible acid rain precursors from natural sources are also important. It is apparent that much work should be done on tracing the release of pollutants, determining their conversion and transport rates, and measuring their deposition rates in order to evaluate the effects of modification of the various source contributions on the acidic precipitation in receptor areas.

In view of these uncertainties, this section of the book focuses on assessing the magnitude and distribution of the precursors of acid rain from anthropogenic and natural sources. A list of acidic or potentially acidifying substances involved in acid rain formation includes:

- sulfur compounds and radicals: sulfur dioxide (SO_2), sulfur trioxide ($SO_3^=$), hydrogen sulfide (H_2S), dimethyl sulfide (($CH_3)_2S$ or DMS), dimethyl disulfide (($CH_3)_2S_2$ or DMDS), carbonyl sulfide (COS), carbon disulfide (CS_2), sulfate ($SO_4^=$), sulfuric acid (H_2SO_4), methyl mercaptan (CH_3SH or MeSH);

- nitrogen compounds and radicals: nitric oxide (NO), dinitrogen oxide (N_2O), nitrogen dioxide (NO_2), nitrite (NO_2^-), nitrate (NO_3^-), nitric acid (HNO_3), ammonium (NH_4^+), ammonia (NH_3); and

- chlorine compounds and radicals: chlorine (Cl^-), hydrochloric acid (HCl).

Precipitation acidity is primarily attributed to the strong mineral acids H_2SO_4 and HNO_3. The immediate precursors of these acids are the man-made and naturally produced gases SO_x (SO_2 and SO_3) and NO_x (NO and NO_2). Natural sources of SO_x and NO_x are generally distributed globally, while anthropogenic emissions tend to be concentrated regionally near population centers.

In addition to the principal sources, other contributors (e.g., ammonia and chlorides) and inhibitors (e.g., ammonia) are important in the formation of acid rain. The role of metal catalysts such as vanadium, manganese, and iron, and the potential synergistic reactions of ozone, carbon dioxide, hydrocarbons, and particulates with the principal causative agents of acid rain need to be analyzed. Many of these pollutants are produced by both stationary and mobile sources. Their roles in smog formation, in the contribution to background acidity in rain, and as sources or sinks of acidic components and precursors of precipitation need further investigation.

The basic source assessment methodology used to map the quantities of the principal sources of acid precipitation is the emission inventory. Data from the 1977 National Emissions Data System (NEDS) have been used in this report as a basis for describing the current trends in the magnitude and distribution of SO_x and NO_x.[1] Although the reliability of the NEDS inventory has been questioned and a number of other inventories are available, the use of NEDS is practical for the following reasons: it has been in operation for several years; it serves as the basis or beginning point for several other inventories; it has a structured updating system; and it is geographically inclusive of the major point and area sources of SO_x and NO_x in the United States. It appears that on a large regional or national scale, the NEDS data base is in general agreement with other inventories.

NATURAL SOURCES OF SO_x AND NO_x

To assess the magnitude and distribution of man-made (anthropogenic) contributions of SO_x and NO_x from individual sources or from source areas, a knowledge of background levels is necessary. A common approach consists of performing a mass balance of the pollutants in a known air volume using approximations of the emission, atmospheric transformation and transport, scavenging, and removal of the pollutant. Budgets, as they are often referred to, have been especially useful in sulfur and nitrogen compound systems by allowing an estimate to be made of natural emission processes, on which few, if any applicable field data are available. Although the magnitude of some natural sources of SO_x and NO_x may be significant, they are globally distributed, whereas man-made emissions tend to be much more concentrated. In polluted urban airsheds, it appears that anthropogenic sources of these pollutants dominate.

Natural Sources of SO_x

The natural sources of SO_x are generally defined to include seaspray containing sulfates from oceans, organic compounds from bacterial decomposition of organic matter, reduction of sulfate in oxygen-depleted waters and soils, volcanoes, and forest fires.[2] A significant natural source of sulfur compounds is the terrestrial and marine biosphere. The releases of hydrogen sulfide (H_2S) and dimethyl sulfide (DMS) may have the greatest impact on the global sulfur cycle.[3] Total terrestrial contributions are estimated at 5 teragrams (Tg)* of sulfur per year,[3] which constitutes a major component of the current total natural and man-made sulfur releases to the atmosphere.[4] Table 2-1 is a summary of available literature values for natural biogenic sulfur emissions.[2]

In most industrialized continental areas, the contribution of volcanic eruptions and fumaroles (holes, usually found in volcanic areas, from which vapors and gases escape) appears to be small in comparison to other natural sulfur emitters. Granat et al.[3] cite an average estimate of elemental sulfur emitted to the atmosphere from volcanic activity to be 3 Tg per year.

Although global and regional sulfur budgets have been estimated, few reliable measurements exist. The numbers for natural emissions have been lowered steadily from the early estimates shown in Table 2-2[11] to about 35 Tg/yr.[3] The fraction produced by man-made contributions has been constantly revised with, in general, an increasing anthropogenic estimated input. Robinson and Robbins[16] have noted that man-made emissions of sulfur compounds have increased fivefold between 1900 and 1965. Whelpdale[17] has summarized recent changes, noting that in the 1960s, the man-made fraction was estimated at about 15 percent; whereas in the 1970s, the contribution had risen to nearly 35 percent. The latest reported estimate[3] attributes approximately two-thirds of all the sulfur emissions to man's activities.

These estimates of natural SO_x emissions indicate that, although it can be highly variable, the natural-emission component of the sulfur cycle is sizable relative to anthropogenic emissions on a global basis. This is not the case, however, in eastern North America. Galloway and Whelpdale[2] have prepared detailed sulfur budgets for Eastern North America (see Table 2-3) showing that man-made emissions account for over 90 percent of the total emissions in the region. Natural emissions account for approximately 4 percent of the total, with inflow from outside the region contributing the remainder. On the basis of other research conducted in the United States and Europe.[18] natural processes (emissions from terrestrial activities, marine processes in coastal zones, and sea-salt advection) were found to contribute about 10 percent to the total SO_x emitted in the eastern United States, which, again, is well below the 35 percent global estimate. Similar conclusions were also reached for Europe.[18]

Natural Sources of NO_x

Relative to global and regional SO_x emission estimates, the inventories for NO_x and related nitrogen compounds are much less certain, particularly for the fraction produced by natural sources. Global fluxes of nitrogen compounds are based largely on extrapolation of experimentally determined, small-scale emission factors, or use of mass balances to obtain crude estimates for

---

*One teragram (Tg) = 10^{12} grams (g) = 10^9 kilograms (kg) = 10^6 metric tons (tonnes,t) = 1.1×10^6 short tons.

TABLE 2-1. LITERATURE VALUES OF NATURAL EMISSIONS OF SULFUR COMPOUNDS[2]

Reference	Source	Compound	Emission rate (mg S $m^{-2}y^{-1}$)	Remarks
Hitchcock, 1975[5]	marine algae vegetation soils	DMS	0.14 3.6 10-33	From anaerobic decay of organic matter; value of 10 is preferred
Adams et al., 1978[6]	soils	H_2S, COS, DMS MeSH, CS_2, DMDS	1.3-7.5	H_2S dominant; daily averages; wet soils; flux was proportional to temperature
	one soil grass (inland) marsh grass	H_2S, COS, MeSH COS, DMS, CS_2 H_2S, COS, CS_2 DMS, DMDS	7.24×10^4 0.4-4.2 $<8 \times 10^2$	tidal area; isolated case
	sea water	COS, DMS, CS_2	4-22	
Maroulis and Bandy, 1977[7]	ocean	DMS	6	Conclusion: marshes, shallow bays, land were not much stronger sources than open ocean
Jaeschke and Haunold, 1977[8]	swamps, tidal flats, anaerobic soils	H_2S	26	
Hansen et al., 1978[9]	sediment of shallow coastal area	H_2S	$1.8-44 \times 10^4$	Probably anaerobic bacterial reduction of SO_4; mainly at night, 2cm H_2O, high temperature limited extent
Liss and Slater, 1974[10]	ocean	DMS	10	Calculated values.
Friend, 1973[11]	land ocean	H_2 sulfide	387 133	Estimated to balance global budget
Granat et al., 1976[3]	anaerobic bacterial reduction of SO_4	H_2S, DMS, others	53	Required to balance budget
Aneja et al., 1978[12]	salt-marsh grass mud flats	DMS, H_2S H_2S, DMS	660 220	Average values; emission rates increase with increasing temperature

TABLE 2-2. GLOBAL EMISSIONS OF SULFUR (Tg/yr)[11]

Item	Eriksson, 1960[13]		Junge, 1963[14]		Robinson and Robbins, 1968[15]		Kellogg et al., 1972[4]		Friend, 1973[11]	
Industry, space heating, and tation (mainly SO_2 and H_2S)	40	(10)[a]	40	(14)	70	(30)	50	(27)	65	(23)
Biological decay, land (H_2S)	110	(29)	70	(24)	68	(29)	90[b]	(49)	58	(20)
Biological decay, ocean (H_2S)	170	(44)	160	(54)	30	(12)	–[b]		48	(17)
Sea spray (sulfates)	40	(10)	–[c]		44	(18)	43	(23)	44	(15)
Volcanoes (SO_2, H_2S, sulfates)	–[d]		–		–		1.5	(1)	2	(1)
Fertilizer application to soil (sulfates)	10	(3)	25	(8)	11	(5)	–		26	(9)
Rock weathering (sulfates)	15	(4)			14	(6)	–		42	(15)
Total (natural sources)	345		225		167		134.5		220	
Total (natural and man-made)	385	(100)	295	(100)	237	(100)	184.5	(100)	285	(100)

[a]Numbers in parenthesis indicate percent of total emissions (natural and man-made).
[b]Kellog et al. estimate a total of 90 Tg/yr of sulfur from decay of land and ocean biota.
[c]Junge's model was for excess sulfur only, so the sea-salt component was not included.
[d]Dashes indicate negligible contribution.

TABLE 2-3. ATMOSPHERIC SULFUR INPUTS FOR EASTERN NORTH AMERICA[a,2]

Term	Magnitude (Tg S yr^{-1}) for Eastern:					
	Canada		U.S.A.		North America	
Man-made emissions	2.1	(47)[b]	14	(91)	16	(93)
Natural emissions,						
sea spray, internal	0.06	(1)	-[c]		0.06	(-)
terrestrial biogenic	0.06	(1)	0.04	(-)	0.1	(1)
marine biogenic	0.2	(4)	0.4	(2)	0.6	(3)
Inflow from oceans	0.04	(1)	0.02	(-)	0.06	(-)
Inflow from west	0.1	(2)	0.4	(2)	0.5	(3)
Inflow to U.S. from Canada	-		0.7	(5)	-	
Inflow to Canada from U.S.	2.0	(44)	-		-	
TOTAL	4.6	(100)	15.6	(100)	17.4	(100)

[a] Area includes Ontario, Quebec, the Atlantic Provinces, the Gulf of St. Lawrence, approximately one-half of the Hudson Bay, and in the United States all area east of 92°W (approximately east of the Mississippi River).

[b] Numbers in parenthesis indicate percent of total emissions (natural and man-made).

[c] Dashes indicate negligible contribution.

unknown sources. Robinson and Robbins[19] estimated the ratio of natural emissions of NO_x from terrestrial and aquatic sources to those from anthropogenic sources to be approximately 7:1. An earlier estimate[16] indicated a higher ratio of 15:1. The downward revision was based on a 55 percent lower estimate of the amount of NO_x contributed by natural sources. By way of comparison, Söderlund and Svensson[20] concluded that the ratio of NO_x emissions from natural versus man-made sources could range from approximately 1:1 to 4 or 5:1. The uncertainties associated with global estimates of natural NO_x emissions along with their contributing sources are highlighted in Table 2-4 (depicting NO_x and related compounds) and Table 2-5 (depicting NO and NO_2).

The data presented suggest that natural terrestrial sources of NO and NO_2 and tropospheric production of NO_x-N by lightning can be significant contributors to the total NO_x background. Although 40 to 108 Tg NO_x-N per year have been estimated to be released from terrestrial sources, most of it is reabsorbed with only 8 to 25 Tg NO_x-N escaping to the troposphere.[20] These authors also suggest that the principal source of gaseous NO_x in terrestrial systems is chemical decomposition of nitrates.

During lightning discharges, tropospheric production of NO_x has been estimated to account for 8 to 40 Tg NO_x-N per year.[24,26,27] If these higher estimates are correct, lightning could account for as much as 50 percent of the total atmospheric production of NO_x on a global basis.[24] This level is comparable to one estimate of global man-made NO_x emissions.[28,29]

TABLE 2-4. ESTIMATES OF GLOBAL EMISSIONS OF OXIDES OF NITROGEN AND RELATED COMPOUNDS (Tg N/yr)

	Delwiche, 1970[21]	Burns and Hardy, 1975[22]	Söderlund and Svensson, 1976[20]	Robinson and Robbins, 1975[19]	Liu et al., 1977[23]	Chameides et al., 1977[24]
NO_x emissions from land to atmosphere	NA	NA	40-108	NA	NA	NA
NO_x emissions from land and sea	NA	NA	NA	210 (NO)	NA	NA
NO_x formed by combustion	NA	15	19	15	NA	NA
NO_x formed by industrial processes	30	30	36	NA	40	NA
Atmospheric NH_3 transformation to NO_x	NA	30	3-8	NA	NA	NA
NH_3 emissions to atmosphere	NA	165 (land and sea)	113-244 (land)	870 (land and sea)	NA	NA
Atmospheric production NO_x by lightning	NA	10	NA	NA	NA	30-40

NA = Not available.

TABLE 2-5. ESTIMATES OF THE GLOBAL EMISSIONS OF NO_x (NO AND NO_2) (Tg N/yr)

	Burns and Hardy, 1975[22]	Söderlund and Svensson, 1976[20]	Robinson and Robbins, 1975[19]	Crutzen and Ehhalt, 1977[25]	Chameides et al., 1977[24]
Natural emissions from land to atmosphere	NA	21-89	NA	NA	NA
Natural emissions from land and sea to atmosphere	NA	NA	210	NA	NA
Tropospheric production by lightning	10	NA	NA	8-40	30-40
Stratospheric production from N_2O	5	0.3	2	NA	NA
Atmospheric production from NH_3	NA	3-8	NA	NA	NA
Production during combustion	15	19	15	NA	NA
Other industrial production	30	36	NA	NA	NA

NA = Not available.

ANTHROPOGENIC SOURCES OF SO_x AND NO_x EMISSIONS

Magnitude and Distribution

Nationwide emissions of sulfur oxides (SO_x) and nitrogen oxides (NO_x) are summarized in this section. These two pollutants, which are the primary precursors in acid rain formation, rank third and fifth, respectively, when compared to national discharges of the other three criteria pollutants (viz., particulates, hydrocarbons, and carbon monoxide). Data from the 1977 NEDS have been used as a basis for analysis throughout this section. The quality of this data base is discussed subsequently in the report.

Total emissions of SO_x and NO_x for 1977 are shown to be 31.5 and 21.7 × 10^6 tons, respectively. Stationary fuel combustion is a major contributor of both pollutants as shown in Figures 2-1 and 2-2. In the case of SO_x, fuel combustion accounts for just over three-fourths of the total, whereas for NO_x, its contribution is about half. The second most important source of SO_x emissions is industrial processes, which account for about 18 percent of the total. Industries pertaining to primary metals (8.9 percent), petroleum (3.1 percent), chemical manufacturing (2.7 percent), and mineral products (2.2 percent) are the major sources. A further breakdown of these industrial process SO_x emissions is possible: copper smelters account for 75 percent of primary metal emissions; process heaters and catalytic cracking operations account for about 60 percent of petroleum industry emissions; sulfuric acid and elemental sulfur production make up 67 percent of the chemical manufacturing category; and cement manufacturing contributes about 90 percent of SO_x emissions in the mineral products industry.

Figure 2-2 indicates that the transportation category is second in importance to fuel combustion, accounting for 43 percent of total NO_x emissions. About two-thirds of the transportation contribution is produced by gasoline combustion, and one-third is produced by diesel fuel combustion.

Although the magnitude of the SO_x and NO_x emission problem is important, of at least equal concern is how these emissions are distributed across the country. Figures 2-3 and 2-4 represent emission density maps for SO_X and NO_X, respectively, wherein emissions per square mile are indicated by progressively darker areas. It is readily apparent that most of the emissions occur in the eastern half of the country. In fact, the 26 states east of the Mississippi River account for 71 percent of the SO_X and 59 percent of the NO_X emissions. Although emissions estimates on a state-by-state basis are more accurate than those made using smaller subdivisions [i.e., counties or Air Quality Control Regions (AQCRs)], emission density maps for SO_X and NO_X (Figures 2-5 and 2-6, respectively)[30] are provided to indicate the "hot spots" within each state.

To provide additional insight into the major contributing states and the important source categories in each state, two tables have been prepared. Table 2-6 provides a list of 19 states that together constitute at least 80 percent of total U.S. SO_X emissions. Total state emissions, percent of U.S.

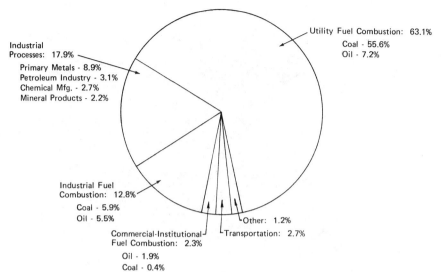

Figure 2-1. Percentage of 1977 national SO_x emissions by source category.

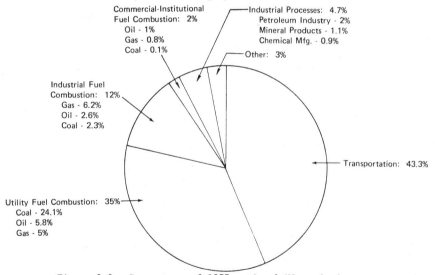

Figure 2-2. Percentage of 1977 national NO_x emissions by source category.

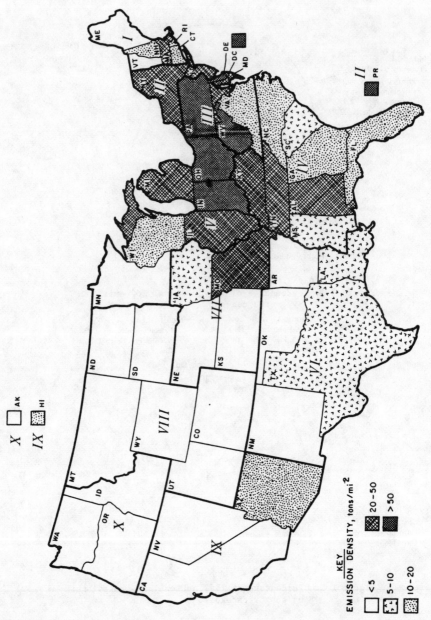

Figure 2-3. Characterization of U.S. SO_x emissions density by state.

Sources 43

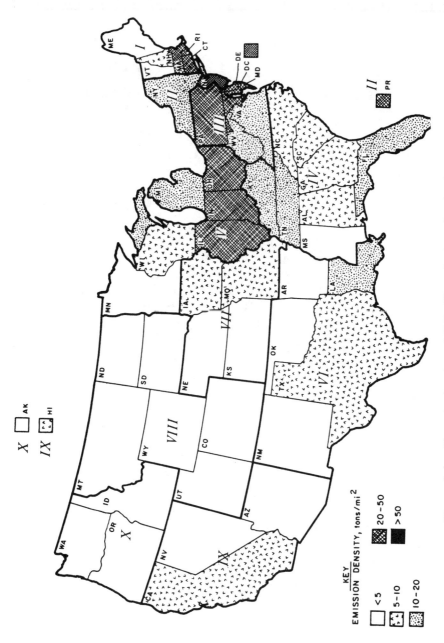

Figure 2-4. Characterization of U.S. NO_x emissions density by state.

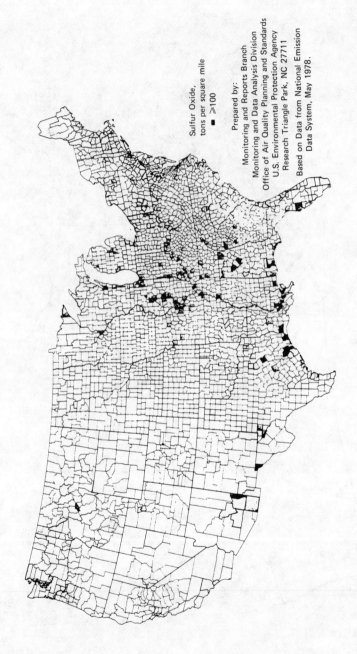

Figure 2-5. SO$_x$ emission density by county.[30]

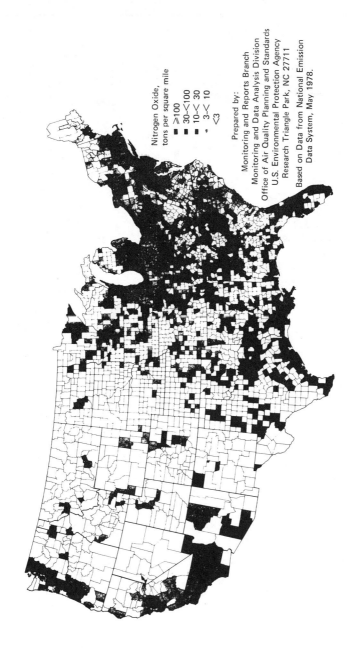

Figure 2-6. NO_x emission density by county.[30]

TABLE 2-6. NATIONAL DISTRIBUTION OF 1977 SO_x EMISSIONS

	State (by rank)	SO_x emissions (10^6 tons)	% U.S.	Major source(s) - % of state
1.	Ohio	3.26	10.3	EG/BC - 82
2.	Pennsylvania	2.5	7.9	EG/BC - 56, IF/PG - 10 IF/BC - 8, PM - 11
3.	Indiana	1.89	6.0	EG/BC - 79
		Subtotal	24.2	
4.	Illinois	1.71	5.4	EG/BC - 78
5.	Kentucky	1.63	5.2	EG/BC - 94
6.	Texas	1.54	4.9	CM - 25, IF - 17 PM - 14, EG/L - 13 PI - 11
7.	Missouri	1.5	4.8	EG/BC - 83
8.	Tennessee	1.28	4.1	EG/BC - 85
9.	Arizona	1.24	3.9	PM - 88
		Subtotal	52.5	
10.	West Virginia	1.23	3.9	EG/BC - 85
11.	Michigan	1.22	3.9	EG/BC - 72, IF/BC - 8.5
12.	Alabama	1.04	3.3	EG/BC - 75 PI - 3, T - 2.5
13.	New York	1.02	3.2	EG/BC - 25, EG/RO - 25 CIF/RO - 13, IF/RO - 11 IF/BC - 9
14.	Florida	0.989	3.1	EG/RO - 39, EG/BC - 37 CM - 6.5
15.	Georgia	0.7	2.2	EG/BC - 73, EG/RO - 9
16.	California	0.675	2.1	PI - 19, EG/RO - 18 T - 16, IF/RO - 16 EG/DO - 9, CM - 7
17.	Wisconsin	0.666	2.1	EG/BC - 70, IF/BC - 15
18.	North Carolina	0.618	2.0	EG/BC - 68, IF/RO - 12
19.	New Mexico	0.576	1.8	PM - 43, EG/BC - 25 PI - 16
		Total	80.1	

NOTE: EG = Electric generation
       IF = Industrial fuel use
       CIF = Commercial/institutional fuel use
       BC = Bituminous coal
       RO = Residual oil
       DO = Distillate oil
       PG = Process gas
       L = Lignite
       PI = Petroleum industry
       PM = Primary metals
       CM = Chemical manufacturing
       T = Transportation

total, and the major source category(s) in each state are indicated. Electric generation by bituminous coal is seen to be a major factor in most states. Table 2-7 shows 25 states that contribute at least 80 percent of the national NO_x emissions. A breakdown for each state is given only between total fuel combustion and transportation because NO_x emissions are fairly evenly split among the numerous fuel combustion subcategories.

Regional Summaries--

Besides looking at emissions on a state basis, it is also of interest to see which sections or regions of the country are important in terms of magnitude and source distribution. The following discussion summarizes emissions by EPA region. (These regions are highlighted on the statewide emission density maps, Figures 2-3 and 2-4). Because stationary fuel combustion and transportation are dominant categories in most states, other source categories are mentioned only if they are significant.

- Region I - (Connecticut, Maine, Massachusetts, New Hampshire, Rhode Island, Vermont). The New England area is not especially noted for heavy industry (the region has many nonpolluting computer, electronic, and service-oriented companies), nor are there many coal-burning utility or industrial facilities. The region contains 5.6 percent of the U.S. population but contributes only 3.4 percent of total NO_x and 2.1 percent of total SO_x.

- Region II - (New Jersey, New York, Puerto Rico, Virgin Islands). This region, consisting of New York, New Jersey, Puerto Rico, and the Virgin Islands, makes up about 13 percent of the U.S. population. The region contributes 5.3 percent and 7.0 percent of U.S. SO_x and NO_x emissions, respectively.

- Region III - (Delaware, District of Columbia, Maryland, Pennsylvania, Virginia, West Virginia). This region, consisting of the five mid-Atlantic states and the District of Columbia, accounts for 11 percent of the population. Total SO_x is 4.7×10^6 tons (15 percent of the U.S. total) and total NO_x is 2.3×10^6 tons or 10.7 percent of total U.S. emissions. Aside from fuel combustion and transportation, primary metals (two zinc smelting complexes) contribute 11 percent of Pennsylvania's SO_x emissions, and cement manufacturing accounts for 11 percent of Maryland's SO_x total.[31]

- Region IV - (Alabama, Florida, Georgia, Kentucky, Mississippi, North Carolina, South Carolina, Tennessee). This region in the southeastern United States, made up of eight states, represents 16.3 percent of the U.S. population. Sulfur oxide emissions are 6.8×10^6 tons (21.6 percent of the U.S. total), whereas NO_x emissions are 3.8×10^6 tons (17.5 percent of U.S. emissions). The only significant industry besides fuel combustion and transportation contributing to emissions is chemical manufacturing, which accounts for about 7 percent of Florida's SO_x total, probably attributable to 14 sulfuric acid plants located in the state.

TABLE 2-7. NATIONAL DISTRIBUTION OF 1977 NO_x EMISSIONS

	State (by rank)	NO_x emissions (10^6 tons)	% U.S.	Percentage of state emissions	
				Fuel combustion	Transportation
1.	Texas	2.12	9.8	59	31.5
2.	California	1.28	5.9	28	60
3.	Illinois	1.27	5.9	61	35
4.	Ohio	1.19	5.5	61	37
		Subtotal	27.1		
5.	Pennsylvania	1.02	4.7	54	42
6.	Indiana	0.96	4.4	68	28
7.	New York	0.91	4.2	50	48
8.	Louisiana	0.80	3.7	58	26
9.	Michigan	0.74	3.4	46	49
10.	Florida	0.68	3.1	41	53
		Subtotal	50.6		
11.	Missouri	0.62	2.9	56	40
12.	Kentucky	0.57	2.6	67	31
13.	Tennessee	0.56	2.6	54	42
14.	North Carolina	0.514	2.4	47	50
15.	Alabama	0.511	2.4	53	41
16.	Georgia	0.472	2.2	38	57
17.	West Virginia	0.471	2.2	79	18
18.	New Jersey	0.45	2.1	37	58
19.	Wisconsin	0.44	2.0	46	42
20.	Virginia	0.42	1.9	39	57
21.	Kansas	0.35	1.6	53	41
		Subtotal	75.5		
22.	Minnesota	0.34	1.6	39	58
23.	Washington	0.31	1.4	32	57
24.	Oklahoma	0.306	1.4	42	53
25.	Maryland	0.305	1.4	39	55
		Total	81.3		

- **Region V** - (Illinois, Indiana, Michigan, Minnesota, Ohio, Wisconsin). Region V has the "distinction" of leading, in both SO_x and NO_x emissions. It contains more people than any other region (∼21 percent of the U.S. population), and it accounts for 29 percent of total SO_x and 23 percent of total NO_x emissions. Stationary fuel combustion is by far the most important source of SO_x emissions in the region with the following percent contribution from each state in the region:

  - Illinois   - 91,
  - Indiana   - 96,
  - Michigan   - 87,
  - Minnesota - 86,
  - Ohio   - 96, and
  - Wisconsin - 92.

- **Region VI** - (Arkansas, Louisiana, New Mexico, Oklahoma, Texas). The five south-central states in this region represent about 10 percent of the U.S. population and account for about 9 percent of total SO_x and 17 percent of total NO_x. The high NO_x contribution is primarily from Texas, which leads the country in this category, contributing 10 percent of the U.S. total. Sulfur oxide emission sources other than fuel combustion that are important in this region are: Texas - 39 petroleum refineries, 14 sulfuric acid plants, 1 copper and 4 zinc smelters; New Mexico - 3 petroleum refineries and 1 copper smelter; Louisiana - 11 petroleum refineries.

- **Region VII** - (Iowa, Kansas, Missouri, Nebraska). These four states in the central United States account for 5.3 percent of the population, 6.7 percent of SO_x emissions, and 6.6 percent of NO_x emissions.

- **Region VIII** - (Colorado, Montana, North Dakota, South Dakota, Utah, Wyoming). The six mountain states in the north-central United States contribute only 2.9 percent of the population, 2.9 percent of SO_x emissions, and 3.7 percent of NO_x emissions.

- **Region IX** - (Arizona, California, Hawaii, Nevada, Guam, American Samoa). Four states and the territories of Guam and American Samoa make up this region, which represents almost 12 percent of the U.S. population. This region contributes 8.1 percent of total NO_x emissions and 7.3 percent of total SO_x emissions. Other than fuel combustion, large emissions of SO_x are produced by 8 copper smelters in Arizona and 30 petroleum refineries in California. Also, 60 percent of NO_x emissions in California are produced by transportation.

- **Region X** - (Alaska, Idaho, Oregon, Washington). The three northwestern states of Idaho, Oregon, and Washington and the state of Alaska constitute Region X, which accounts for 3.3 percent of the population, but only 1.3 percent of SO_x emissions and 3.1 percent of national NO_x emissions.

In an attempt to correlate SO_x and NO_x emissions to population, linear regression analyses were performed on both a state and regional basis using NEDS data. Correlation coefficients, denoted by r (with +1 being a perfect linear correlation), obtained were as follows:

- SO_x, regional, r = + 0.884;
- NO_x, regional, r = + 0.876;
- SO_x, state, r = + 0.568; and
- NO_x, state, r = + 0.826.

Regional emissions appear to be fairly strongly dependent on population, and it is expected that a regional correlation would be stronger than that made on a state basis for SO_x because utilities in one state often supply electricity to neighboring states.

Historical Data--
Nationwide estimates of SO_x and NO_x are available from the EPA emission estimates[30,32] for 1940 through 1977. These data are shown graphically in Figures 2-7 and 2-8 for SO_x and NO_x, respectively. Although the data from 1970-1977 are more reliable than those for the years 1940, 1950, and 1960, both pollutants show a significant increase from 1960 to 1970. Between 1940 and 1960, the estimated local U.S. emissions of SO_x to the atmosphere increased very little, from 21.9 million short tons per year (21.9 x 10^6 tons/yr) to 24.1 x 10^6 tons/yr, whereas NO_x increased from 6.7 x 10^6 tons/yr to 11.5 x 10^6 tons/yr. In the period from 1960 to 1976, SO_x reached its peak in 1970 with 32.1 x 10^6 tons/yr, viz., a 46 percent increase from 1940; whereas NO_x rose to 25.1 tons/yr in 1973, representing a 273 percent increase from 1940 and a 117 percent increase from 1960.[33]

The large growth in utility emissions was compensated in part by reduction in industrial and residential/commercial emissions, produced by switches from coal to oil and gas in the 1940s and 1950s. Most utility emissions are from coal, reflecting the impact of new sources equipped with flue gas desulfurization (FGD) systems[33] in the 1970's, and, to a lesser extent, the impact of mandated fuel switching (resulting from the Energy Supply and Environmental Control Act--ESECA and the Power Industry Fuel Use Act--PIFUA). Transportation, industrial, and utility sources contributed most to increased NO_x emissions.

Projections--
Figures 2-9 and 2-10 present the national trends for SO_x and NO_x discussed above along with projections to the year 2000. The methodology and assumptions for these scenarios can be found in the Regional Issue and Assessment (RIA) Program report.[34] These emission estimates are based on several future energy scenarios that are currently being analyzed, and reviewed at Brookhaven National

Sources 51

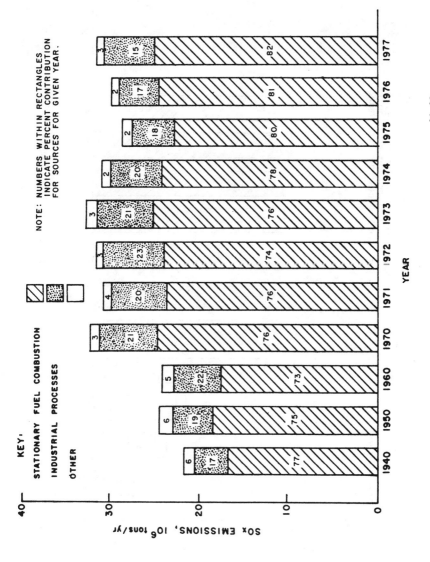

Figure 2-7. Trends in national SO_x emissions since 1940.[30,32]

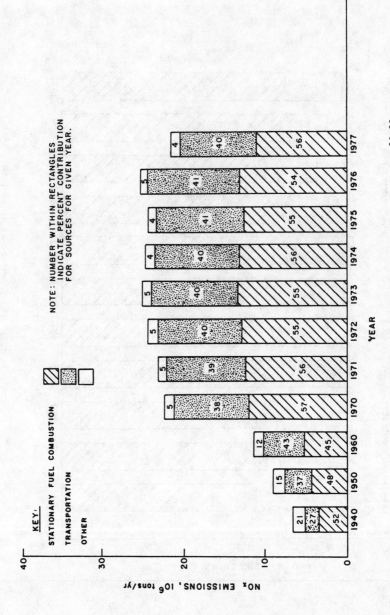

Figure 2-8. Trends in national NO_x emissions since 1940.[30,32]

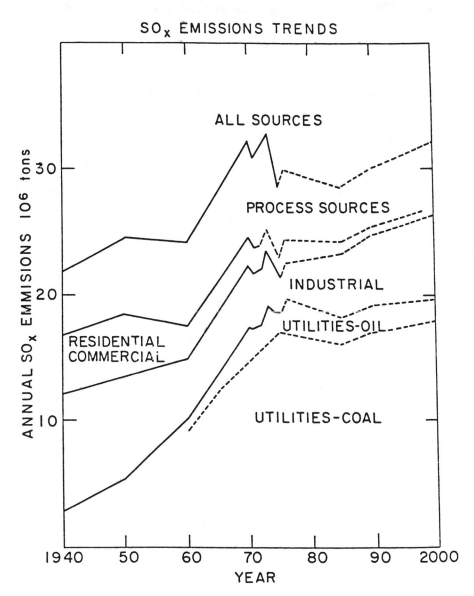

Figure 2-9. SO$_x$ emissions trends for the U.S., 1940-2000.[33]

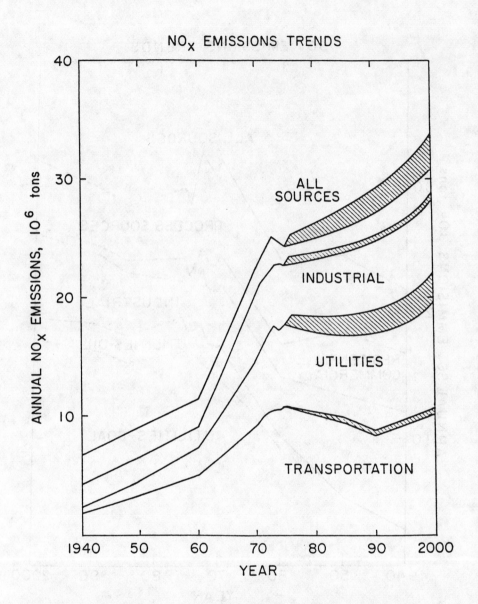

Figure 2-10. NO$_x$ emissions trends for the U.S., 1940-2000.[33]

Laboratory under the sponsorship of DOEs Office of Technology Impacts (now the Office of Environmental Assessments).[33] These scenarios are defined in the Second National Energy Plan (NEP II), with future year data adjusted to correspond to the 1975 EPA data to present continuous trends. The shaded areas on Figure 2-10 represent the range of scenarios defined in NEP II for NO_x. Transportation will continue to be a major source of NO_x, although the dip noted in Figure 2-10 between 1975 and 1990 corresponding to the imposition of emission controls, may not be realized due to waivers being given to automobile manufacturers. Utility and industrial emissions will continue to grow for NO_x, but will remain relatively constant for SO_x. For SO_x, variations in the NEP II scenarios are much larger because of the uncertainties of retrofit controls on existing power plants, and therefore the emissions shown probably represent an upper limit.[33]

Although these figures give a complete picture of national trends, a regional-scale analysis emphasizing planned coal-fired electric generating capacity (an important contributor to SO_x and NO_x production) will aid in assessing environmental impacts. Table 2-8 shows current and planned increases in electric generating capacity by EPA Federal region for coal and oil over the period 1979 through 1989 based on the results of a survey conducted by the National Coal Association.[35] Note that sizable increases (greater than 25 percent) in coal burning occur in all but three northeastern regions (Regions I, II, and III). Historically, the Northeast's contribution to total U.S. SO_x and NO_x has been decreasing (based on the NEP II scenarios mentioned above); viz., 30 percent of the SO_x in 1970, 25 percent in 1975, and 18 percent projected in the year 2000; and 20 percent for NO_x in 1970, 17 percent in 1975, and 14 percent in the year 2000.[33] In view of these facts, it appears that the extent of interregional transfer of airborne sulfur and nitrogen oxides produced by coal burning in regions lying to the west may be an important environmental issue for the Northeast.

Local Sources--

Although long-range transport of SO_x and NO_x from coal-fired power plants appears to be a significant factor for the Northeast, local sources of SO_x and NO_x may also contribute to the acid rain phenomenon in this region as well as in other parts of the country. It has been suggested (as will be discussed in more detail under <u>Factors Affecting Source Emissions</u> -- Combustion Variables) that local oil-fired sources are especially suspect for two reasons:

1. Burning of both distillate and residual oil produces large quantities of primary sulfates, which are formed in the furnace and need not undergo chemical reaction to participate in formation of acid rain. Small residential and commercial boilers emit a much higher percentage of primary sulfates than do utility oil-fired boilers.

2. Burning of residual oil releases large quantities of finely divided catalytic materials, such as vanadium and carbon, which can catalyze the transformation of sulfur dioxide to sulfate as these substances remain suspended in the atmosphere.

TABLE 2-8. CURRENT AND PLANNED INCREASES IN U.S. ELECTRIC GENERATING CAPACITY BY EPA REGION, 1979-1989[35]

EPA region	Current (1978) capacity (MW)	Planned increase (MW)			Percent increase		
		Coal	Oil	Total	Coal	Oil	Total
I	11,984	600	-	600	5	-	5
II	23,580	1,550	850	2,400	6	3.5	10
III	49,948	9,547	1,820	11,367	19.1	3.7	22.8
IV	83,545	31,522	2,222	33,744	37.7	2.7	40.4
V	84,875	21,730	1,350	23,080	25.6	1.6	27.2
VI	73,771	37,886	480	38,366	51.4	0.6	52
VII	22,725	11,870	-	11,870	52.2	-	52.2
VIII	12,274	15,770	-	15,770	128	-	128
IX	29,386	7,896	-	7,896	26.9	-	26.9
X	1,329	530	-	530	40	-	40
Totals	393,417	138,901	6,722	145,623	35.3	1.7	37

In addition to the Northeast, acidification of lakes in Florida and California may be partially attributable to local source contributions.[131] In California, for example, there are a significant number of petroleum refineries and nearly all stationary combustion facilities burn oil or natural gas. In Florida, the majority of the fuel burning facilities use residual oil and there is probably an important contribution from the state's sulfuric acid and phosphate mining industries. On the other hand, researchers in Florida have noted that the more acidic rain is associated with air masses from the north. Both of Florida's northern bordering states are large coal-burning states; 92 percent of Georgia's electric generation is by coal and 76 percent of the state's SO_x emissions are due to coal burning. In Alabama, the figures are 98 and 80 percent, respectively. These data may suggest that long-range transport is a factor in Florida's acid rain problem (see Section 3). In both Florida and California, it should be noted that emissions of NO_x are significant--especially from mobile transportation sources--and may also contribute to local rain acidification.

Canadian Emission Inventory--

The data cited in this section for SO_x and NO_x emissions in Canada are primarily from the work of Voldner, Shah, and Welpdale.[36] This inventory has been developed by the Canadian Department of the Environment as part of its Long-Range Transport of Air Pollutants (LRTAP) Program. The inventory consists of point and area sources mainly for 1974,[37] but in some cases updated to 1976 and 1977.

Total annual emissions are given as 6.5×10^6 short tons for SO_2 and 2.2×10^6 short tons for NO_2. (These values are 21 and 10 percent, respectively, of SO_x and NO_x emissions in the United States). Emission density maps for each of these pollutants are shown in Figures 2-11 and 2-12, where the numbers cited represent ktonnes/yr. Point sources contribute 80 percent of the total SO_2 emissions, and area sources such as transportation and fuel combustion account for 85 percent of total NO_2 emissions. Ontario, where 90 percent of the sources are classified as point sources, contributes 40 percent of all SO_2 emissions. These sources are concentrated in several areas. For example, the Sudbury smelting complex in Ontario combined with the utility sector account for 80 percent of the point source emissions, and in Alberta, the Oil Sands development and natural gas fields account for 90 percent of point source emissions. Quebec, where 71 percent of the sources are classified as point sources, accounts for 23 percent of total SO_2 emissions.

Although data are currently unavailable for long-term projections of Canadian sulfur emissions, they should be similar to trends in the United States.[38] Canadian smelter emissions are expected by the end of the century to produce SO_2 emissions ranging from current levels to 60 percent of these levels as older plants are phased out and replaced by newer and better controlled facilities. As with SO_2 emissions, NO_2 projections are expected to follow U.S. trends with increases primarily in the utility sector.[38]

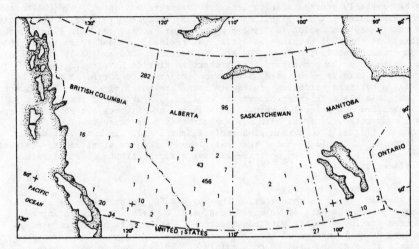

a. SO$_2$ emissions in western Canada from all sources (10^3 MT/yr).

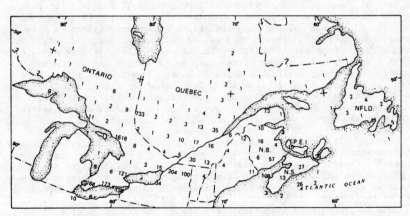

b. SO$_2$ emissions in eastern Canada from all sources (10^3 MT/yr).

Figure 2-11. Total SO$_2$ emissions in Canada.[36]

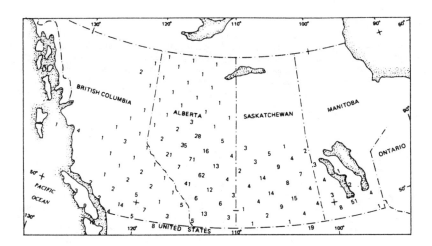

a. NO_2 emissions in western Canada from all sources (10^3 MT/yr).

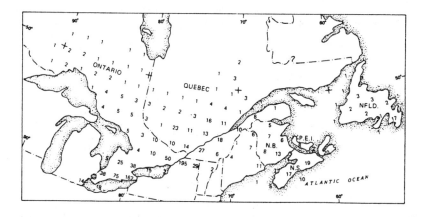

b. NO_2 emissions in eastern Canada from all sources (10^3 MT/yr).

Figure 2-12. Total NO_2 emissions in Canada.[36]

Factors Affecting Source Emissions

Combustion Variables--
    The amount of uncontrolled SO_x emissions from a utility or industrial boiler depends on the amount and sulfur content of the fuel burned, the type of boiler, and other chemical and physical properties of the fuel. In the combustion of bituminous and anthracite coals, approximately 95 percent of the coal sulfur is emitted as gaseous SO_2 and SO_3; conversion of sulfur compounds in these fuels to SO_2 is 90 to 100 percent complete.[39,40] The remaining coal sulfur may be emitted as particulate sulfates or may combine with the slag or ash in the furnace and be disposed of as solid waste. The degree of SO_3 formation, usually 1 to 2 percent of total SO_x, depends on combustion conditions, with, in general, a leaner fuel mixture forming more SO_3.[41] Other factors that will affect SO_3 formation are boiler age, boiler design, and method of firing.

    Sulfur retention in anthracite and bituminous coal ash is generally low, ranging from less than 1 percent at high ashing temperatures (1200°C) to 15 percent at low ashing temperatures (750°C).[42] There is some evidence that sulfur in coal of high alkaline ash content tends to concentrate in the ash during combustion. A high-sodium lignite may retain over 60 percent of the available sulfur in the boiler,[43,44] whereas a low-sodium lignite may contain less than 10 percent of the available sulfur in the boiler ash.[45,46]

    Other studies have shown that excess boiler oxygen enhances the emission of primary sulfates from both oil-fired and coal-fired boilers.[47,48] In the presence of excess oxygen, SO_2 is oxidized to SO_3 which then hydrolyzes to H_2SO_4. For a given fuel sulfur content, oil-fired units emit more SO_3 than coal-fired units.[47] In general, industrial boilers (usually stoker-fired) use higher excess oxygen levels than do utility boilers (predominantly pulverized coal-fired). Another source of primary sulfate emissions is frequent soot-blowing operations on boilers with catalytically active sulfate deposits.

    Preliminary studies on emissions characterization of fossil-fuel burning sources have indicated substantially greater sulfate emissions from high-vanadium content residual oils as compared to pulverized coal.[49] The higher flame temperature of oil serves to exacerbate the formation of SO_3, H_2SO_4, and particulate sulfates, while the presence of vanadium catalyzes the formation of SO_3 in the combustion process.[47] As noted above, excess oxygen also serves to enhance primary sulfate emissions from oil-fired boilers. However, at this time, the precise mechanism of sulfate formation from oil-firing is not well defined.[49]

    Emissions of NO_x from coal-fired power plants depend on the type of boiler and the manner in which it is operated. One of the most important variables is excess air and how it relates to temperature.[50] For a given furnace temperature distribution, the quantity of NO_x formed decreases as the excess air decreases. This is because of the reduced oxygen concentration in the high temperature flame zone in the furnace where the NO_x is formed. Reductions in the oxygen content decrease emissions of both fuel NO_x (generated by the oxidation of fuel contaminants that contain nitrogen) and thermal NO_x (generated by the reaction of atmospheric oxygen and nitrogen). Reductions in temperature, however, produce significant reductions only in thermal NO_x.

The type of boiler is also important in determining expected NO_x emissions. Cyclone boilers produce greater amounts of NO_x than pulverized coal-fired boilers, and these emissions are greater than those from stoker boilers.[51] This is reflected in EPA emission factors[52] for each of the three boiler types as shown below.

Boiler type	Emission factor (lb NO_2/ton coal)
Pulverized	18
Stoker	15
Cyclone	55

These trends are consistent with the fact that cyclones have the highest heat release rates and furnace temperatures and that stokers have the lowest.

Emissions of NO_x from wet bottom boilers are higher than those from dry bottom boilers, which operate at lower temperatures. Lowest temperatures are found in tangentially fired units.[50] An analysis of 53 coal-fired utility boilers showed no apparent correlation between NO_x emissions and boiler size.[53]

With regard to oil firing, an analysis of NO_x emissions from 31 oil-fired utility boilers,[54,55] showed that for tangentially-fired units the average NO_x emissions were 0.27 lb/10^6 Btu, while for all other types it was 0.55 lb/10^6 Btu. Comparison of these rates with EPA emission factors shows that the difference between emission factors for tangentially-fired units is 20 percent; whereas for other types it was 25 percent. A similar comparison for 28 gas-fired boilers[54,55] shows a 7 percent difference between average measured and EPA emission factors for tangentially-fired units; and 19 percent for other units.

Although the age of a boiler may affect the release of SO_x and NO_x somewhat (due to decreased thermal transfer over time), the allowable emissions for boilers predating New Source Performance Standards (NSPS) and State Implementation Plans (SIPs) will have a greater impact on total emissions. Older utility boilers are generally located in urban population centers and have shorter stacks, hence the local impact may be greater than from combustion units located in remote areas and having taller stacks. Individual SIPs sometimes make this distinction and disallow the older plants from emitting disproportionately higher emissions. In any event, the retirement of these older units may be important to the acid rain issue, since newer boilers replacing the older units over time, would be required to meet more stringent emission control levels.

Control Technology—
Emission control equipment may have a significant impact on the composition and concentration of SO_x from coal-fired boilers. In addition to low sulfur coal-firing, coal cleaning, dry sorbent SO_x removal, and fluidized-bed combustion (FBC) technologies are also receiving considerable attention for reducing SO_x and, in the case of FBC, also NO_x emissions. Flue gas desulfurization systems may be designed to remove SO_2 with greater than 90 percent removal efficiency.[56] Although electrostatic precipitators, used to control particulate emissions from boilers using FGD systems, usually have little effect on overall SO_2 emissions, they may help to reduce total sulfate emissions by 50 percent or more.[47] However, conditions in hot side electrostatic precipitators may promote sulfate formation. Arcing of corona discharges may cause localized hot spots at high temperatures as well as produce ozone, which can oxidize SO_2 to SO_3.[57] Also, because sulfate particles are generally submicron in size, they are generally not efficiently collected by electrostatic precipitators, thereby leading to sulfate enrichment of fine particulate matter in the outlet stream.[47,48,57]

Primary sulfate removal from FGD systems appears to be very dependent on design practices. Although some tests on a wet scrubber system exhibited primary sulfate removal efficiencies of up to 30 percent,[58] other results in the same study showed increases in primary sulfate in the scrubber outlet stream. These increases may be produced by gaseous H_2SO_4 penetration of the scrubber demisters forming H_2SO_4 aerosol and by scrubber liquor reentrainment.

Fuel additives usually containing magnesium have been used to effectively reduce primary sulfate emissions from oil-fired boilers.[47] These additives are selected for their ability to minimize the formation of SO_3 by reacting with or absorbing SO_3 and H_2SO_4. The reaction products are then retained in the bottom ash in the boiler or combine with fly ash which may then be collected by an electrostatic precipitator.

Because reduction in temperature in the flame zone of a boiler can produce significant reduction in thermal NO_x, several methods have been developed and used. These methods include: injection of cooled combustion products, steam, or water into the flame zone; reduction of the temperature to which combustion air is preheated; extraction of heat from the flame zone; and reduced furnace load.[53] Several methods have also been used to reduce oxygen content in the flame zone. These methods include the use of low excess air firing, staged combustion, flue-gas recirculation, reduced furnace load, water or steam injection, and reduced air preheat. The above oxygen reduction techniques have been successfully demonstrated on utility boilers, although the latter two methods also reduce thermal efficiency. Using combinations of the techniques listed above, an average reduction of NO_x emissions of 37 percent has been achieved for coal-fired utility boilers, 48 percent for oil-fired boilers, and 60 percent for gas-fired boilers.[51]

Other Variables—
Variations in NO_x emissions can be expected because of seasonal variation in power production from fossil-fuel generating plants. On a nationwide basis, the variation is estimated at 15 percent; greatest production is in the summer

and least is in the spring.[59] However, greater variation as well as different seasonal patterns have been reported for different areas of the country.[60] Factors that influence seasonal variabilities in mobile source emission of NO_x include temperature dependencies of emissions per vehicle mile traveled (viz., a 35 percent reduction as temperature increases from 20 to 90°F)[61] and changes in seasonal vehicle miles traveled (viz., nationwide, production is 18 percent higher in summer than in winter).[62] Also, the potential impact of diurnal variations associated with motor vehicle traffic on ambient air quality are also important. Representing mobile sources by annual emissions data may underestimate their potential for producing short-term (peak) concentrations.

Variations in SO_x emissions can also be expected due to seasonal variation of major sources. Seasonal average grid SO_x emissions from the Sulfate Regional Experiment (SURE) program (including utility, industrial, commercial, residential, and transportation sources) show a peak in SO_2 during the winter months.[63] During the summer, high temperature and atmospheric moisture content and the persistence of anticyclones contribute to sulfate formation and the regional accumulation of sulfates and other air pollutants. This phenomenon is discussed in detail in Section 3.

Of the industrial process sources discussed earlier, the primary metals industries (principally copper, lead, and zinc) are potentially the largest SO_x emitters. Factors affecting SO_2 emissions from smelters include the quantity of raw ore processed, the ore sulfur content, the process configuration, and the degree of sulfur removal and control for each process step. The major portion of SO_2 is formed in the roasting, smelting, sintering, and converting processes.[64]

Quality of Data Base

The emission inventory that serves as the basis for estimates of the magnitude and distribution of SO_x and NO_x described above, is NEDS,[1] which is operated through the EPA Office of Air Quality Planning and Standards (OAQPS). This inventory contains point and area source (including mobile) emissions data for the criteria pollutants that can be assembled according to individual facilities, AQCRs, states, and the nation for a variety of source categories. Although the reliability of the NEDS inventory has been questioned and a number of other inventories are available, the use of NEDS is practical for the following reasons: it has been in operation for several years; it serves as the basis or beginning point for several other inventories; it has a structured updating system; and it is geographically inclusive of the source categories of interest in the present study.

The NEDS inventory contains information for both point and area sources. Point sources are defined as facilities that emit 100 short tons/year or more of any of the criteria pollutants, although smaller point sources sometimes reported by the states are included here. All other sources (including mobile sources and the unreported point sources less than 100 tons/year) are defined as area sources. Point source data were originally obtained from state air pollution control agencies. These agencies are required to submit revised data for certain sources on an annual basis. Information reported for point sources includes the type of process, operating rates and schedules, pollution control device and efficiency, stack parameters, location data, fuel type and ash, sulfur and heat content, and emission rates. Emissions data consist of stack measurements, engineering calculations, emission factors, or, in some cases, a guessed estimate.

The NEDS area source file contains data on mobile sources, small stationary sources, and miscellaneous sources such as forest fires and retail gasoline distribution. States are not required to submit area source data. Rather, the area source inventory is developed from data published by other Federal agencies and from validated data collected from state agencies. Area source data are updated annually by EPA's National Air Data Branch (NADB).

In concept, the NEDS data base should be a good indicator of emissions for most geographical areas. However, in reality, the data base may contain inaccuracies. Some of the factors that may contribute to the uncertainty are presented below:[1]

- unclassifiable source type, resulting in missing emissions or guessed estimates,

- mistakes and biases in the reported data,

- lack of complete and timely reporting by state agencies,

- emission factors of low reliability and/or inappropriate application,

- missing industrial process emissions from facilities that emit less than 100 tons/year, and

- inaccuracies in the procedures used to compile the NEDS area source data.

Inaccuracies introduced by the first item are considered to be minor in comparison to other sources of error.[1] Major mistakes in the data are investigated using edit programs or by reviewing NEDS output reports at NADB. However, other investigators have encountered such errors when using NEDS to develop other emission inventories.[65,66] Inaccuracies may exist in the area source data for small geographical areas,[1] presumably caused by the methods used to allocate statewide data to the county level. Although the area sources should constitute the difference between the total emissions and point source emissions, some point sources that emit less than 100 tons/year may not be included. In small geographical areas where these sources constitute a significant portion of the total emissions, the NEDS emission inventory will be biased.[1]

Emission factors introduce uncertainty in the emission inventory, particularly as the region of interest narrows. An emission factors is simply the mass of emissions per unit of production or consumption. If a number of these are averaged for a particular type of operation (e.g., an oil-fired boiler) the result is an emission factor that is generally representative of the operation. EPA has compiled emission factors for numerous source types based on the results of measurements and engineering analyses.[52] When applied to many sources of the same type, these emission factors often provide fairly accurate estimates of emissions. When applied to geographical areas where the distribution of sources is different form the national or industry average, the accuracy of the estimate decreases.

EPA has prepared nationwide emission estimates and compared them with emission totals derived from NEDS.[1] Nationwide fuel use, industrial production and other appropriate data were used to determine activity levels for each source category. National average emission factors and pollution control device efficiencies were also used. This methodology is advantageous because all sources are taken into account. It has the disadvantages that measured emission rates from individual sources are not used and that the effect of location (important in mobile source emissions) is not taken into account. A comparison of EPA estimated emissions and NEDS emissions reveals that the totals for SO_x and NO_x agree to within approximately 5 and 15 percent, respectively, although for some source categories, the deviation exceeds 20 percent.[1]

Several emission inventories have been cited in the literature. Dykema and Kemp[61] have developed a nationwide inventory of NO_x, particulate, hydrocarbon, and CO emissions from stationary combustion related sources. An inventory of SO_x, NO_x, particulate, and hydrocarbon emissions for the eastern United States and southeastern Canada was developed for the Sulfate Regional Experiment (SURE).[65] Work is continuing at Brookhaven National Laboratories to develop an emission inventory for the northeast United States as part of the MAP3S program.[36] To aid long-range transport studies, Clark[67] has developed an inventory of Canadian and U.S. emissions east of the Rocky Mountains. Galloway and Whelpdale[2] used an inventory of sulfur emissions to develop a sulfur budget for eastern North America. The U.S.-Canada Research Group on the LRTAP program has cited an inventory composed of the U.S. portion of the SURE inventory and an updated Canadian inventory.[38] EPA has initiated efforts to develop an inventory for the Northeast Corridor, which involves a careful review and update of each state's emission inventory and use of NEDS data for a few states. The NEDS and SURE data bases were used as an information source for most of these studies.

Although it does not include the total United States, the SURE emission inventory is very comprehensive (particularly from a modeling viewpoint) and possesses the following features:

- it encompasses the eastern United States, and southern Quebec and Ontario;

- the criteria pollutants are divided into component species (SO_2, sulfate, NO, NO_2, and hydrocarbons by reactivity);

- seasonal and diurnal variations in emissions are included; and

- hourly emissions of SO_x from the larger power plants are included for selected months.

The SURE inventory was partially derived from NEDS and expanded to include the features noted above. A fairly recent development, it is beginning to be adopted as the basis for regional and individual point source inventory studies as well as long-range transport studies.

The U.S. portion of the SURE emissions inventory has been compared with the NEDS emission inventory for the same region.[1,65,68] This comparison shows that for total SO_x and NO_x emissions, the two inventories differ by less than 10 percent. Comparison of the SURE inventory with Whelpdale and Galloway's sulfur budget for the eastern United States[2] shows that sulfur emissions differ by less than 10 percent.[69]

Attempts have been made to evaluate the accuracy of several of the inventories mentioned above. Accuracy, as defined here, is simply an aggregate estimate of the uncertainties attached to individual reported and calculated data contained in the inventory. Klemm and Brennan[65] have reported an accuracy (uncertainty) of 17 percent for SO_2 and NO_x emissions in the SURE region. Dykema and Kemp indicate an accuracy of about 11 percent for their nationwide inventory of NO_x emissions from stationary source combustion processes.[66] Galloway and Whelpdale report an accuracy of 15 percent for the sulfur inventory they used.[2] Note that these accuracy estimates relate to regional and national annual emissions; they do not describe seasonal variations or the accuracy at the state or small geographical area level.

In summary, several other emission data bases exist for emission inventory analysis. Some are based on the NEDS and SURE inventories with various degrees of review, updating, and additions. It is difficult to determine which is the best inventory, although those which are continually updated are presumably more accurate. It appears that on a large regional or national scale, the NEDS data base is in general agreement with other inventories. In examining such an emission inventory for smaller component regions, however, the emission estimates are subject to higher uncertainties. These greater uncertainties, which exist for NEDS as well as other emission inventories, arise from the manner in which the inventories are developed (e.g., use of nationwide emission factors for some sources which may not apply when examining smaller individual component regions). Still, the NEDS inventory remains as one of the most comprehensive for describing the magnitude and distribution of emissions on a national scale.

OTHER SOURCES

Other Sources Affecting Acid Rain Formation

In addition to the principal sources, ammonia and chlorides are also important in the formation of acid rain. These are discussed in the following sections.

Ammonia--
Ammonia (NH_3) is a colorless gas with a pungent odor which is readily soluble in water, forming ammonium (NH_4^+) and hydroxyl (OH^-) ions. Liljestrand and Morgan[65] suggested that the pH of rain water is controlled by the interaction of bases (NH_3, metal carbonates, oxides) and acids (HNO_3, H_2SO_4). Ammonia is generally an alkaline vapor with the capability of neutralizing either sulfuric or nitric acid in the atmosphere.[71] The presence of ammonium has been found to increase the pH of rain and snow, indicating the alkaline nature of the NH_4^+ ion.[72]

The ammonia in the atmosphere can react directly with sulfuric acid to form ammonium sulfate [$(NH_4)_2SO_4$] particles, a component in smog formation.[73] In gas phase reactions, Scott and Lamb[73] have shown that free SO_2 is removed by ammonia, which thereby decreases the available SO_2 that may otherwise be transformed into H_2SO_4. The National Center for Atmospheric Research (NCAR)[74] showed that the interaction of gaseous NH_3 and SO_2 in the presence of water produces ammonium sulfite and ammonium sulfate, thus suggesting that gaseous ammonia could help neutralize acid precipitation in the atmosphere.

Although extensive studies on the neutralization of nitric acid (HNO_3) by ammonia in the atmosphere are not available, an equilibrium relationship beweeen NH_3, HNO_3 and NH_4NO_3 does exist.[75] In atmospheric reactions with nitrogen oxides, ammonia acts both as a neutralizer of nitric acid and as a promoter of the precursor components of nitric acid. Ammonia may react with hydroxyl radicals to form NO_x at an estimated rate of 20 to 40 x 10^6 metric tons of NO_x-N per year.[24,76] Under varying atmospheric conditions, NO may be catalytically oxidized to nitrites and nitrates.

Most ammonia emissions are released into the atmosphere by natural and biological processes, primarily through the decay and decomposition of organic matter (dead plants, animal and human excreta, etc.)[77,78] Other natural sources include emissions from forest fires and volatilization from land and ocean masses.[79] Decayed organic matter is converted into amino acids through proteolysis and then into ammonia. The ammonia in turn is oxidized by bacteria to form nitrites and nitrates, which are readily assimilated by plants or micro-organisms to produce necessary proteins.[80] In the overall flux of nitrogen between terrestrial and atmospheric biospheres, the process of ammonification is an essential one. Estimates of the natural global ammonia emissions range between 113 to 224 × 10^6 metric tons of N per year.[20] Table 2-9 lists the natural sources of global ammonia and their emission estimates.

TABLE 2-9. NATURAL SOURCES OF AMMONIA AND ESTIMATED EMISSIONS

Source	Global emission rate (10^6 metric tons/yr)
Biological decay	1054[a]
Volatilized from land and sea	165 as N[c]
	860 as N[d]
Volatilized from land	113 - 244 as N[b]

[a] Wark and Warner (1976).[81]
[b] Söderlund and Svensson (1976).[20]
[c] Burns and Hardy (1975).[22]
[d] Robinson and Robbins (1975).[19]

Anthropogenic sources, shown in Table 2-10, account for a small percentage of the total ammonia emissions. These sources are usually the producers of ammonia or users of ammonia in the manufacture of other materials. Approximately 80 percent of the ammonia manufactured by the United States in 1977 was used to produce fertilizers.[84] The remainder of the NH_3 was incorporated in the production of explosives, animal feed, nitric acid, acrylonitrile and amines. The nitrogen fertilizer can be applied directly as anhydrous ammonia or aqueous ammonia,[82] where the inefficient handling and application of fertilizers can result in substantial ammonia losses. The global loss of NH_3 from the handling and application of ammonia has been estimated at 7×10^6 metric tons per year.[85] The United States uses one-quarter of all fertilizers produced[83] and therefore may have the same proportion of global emissions. Ammonia is also generated as a by-product of making coke from coal, ore refining, and fossil-fuel combustion. Söderlund and Svensson[20] calculated that 4 to 12×10^6 metric tons per year of NH_3 are emitted by coal combustion sources on a global basis.

Urea nitrogen [$CO(NH_2)_2$] may volatilize from feedlots where large quantities of animal urine are generated. The National Research Council[86] has estimated that as much as 50 to 100 percent of urea may be volatized and hydrolyzed into NH_3 and CO_2, causing local odor and nuisance problems. An estimated 2 to 4×10^6 metric tons of ammonia may be generated by feedlots.[20]

While it has been suggested that aqueous ammonia can react with and neutralize the acidity in precipitation, the use of ammonium compounds in fertilizer applications may have an adverse impact on ground water, lakes, and streams, in terms of increasing acidity. Although extensive research information on this topic is scarce, it has been noted that the acifification of soil can result from the application of nitrogen fertilizer (e.g., ammonium sulfates, nitrates, and phosphates).[87,88] It is plausible that

TABLE 2-10. ANTHROPOGENIC SOURCES OF AMMONIA AND ESTIMATED EMISSIONS

Source	U.S. emission rate (10^6 metric tons/yr)	Global emission rate (10^6 metric tons/yr)
Natural gas- and coal-based ammonia production	0.017[a]	
Application of anhydrous NH_3	0.153[a]	
Ammonium nitrate production	0.054[a]	
Sodium carbonate production	0.029[a]	
Diammonium phosphate production	0.013[a]	NA
Ammoniator-granulator	0.009[a]	
Urea production	0.009[a]	
Miscellaneous emissions from fertilizer production	0.002[a]	
Beehive coke oven	0.002[a]	
Coal combustion		4-12[b]
Power generation		1.09[c]
Industrial combustion		1.64[c]
Refinery cracking operation		0.018[c]
Fuel oil combustion		0.73[c]
Natural gas combustion	NA	0.018[c]
Incineration		0.073[c]
Wood combustion		0.054[c]
Forest fires		0.045[c]
Urea hydrolyzing (to NH_3 and CO_2)		2-4[d]
Wastewater treatment	3.63[e]	NA

[a] EPA 600/1-77-054 (1977).[82]
[b] Söderlund and Svensson (1976).[20]
[c] Liptak (ed.) (1979).[71]
[d] NRC (1978).[83]
[e] Wark and Warner (1976).[81]
NA = Not available.

rain water run-off into bodies of water at lower elevations than the fields where fertilizers were originally applied, can lower the pH of lakes and streams. With respect to the acidification of Adirondack lakes, such an event is unlikely, since the few agricultural tracts in that region are at lower elevations than the lakes themselves. In addition, since many lakes that are susceptible to damage from acid deposition (discussed in Section 3) are located in forested areas where fertilizers are not locally used, it is unlikely that fertilizer leachate run-off will be a significant component of acidity in these lakes and streams.

Chlorides--

The rainwater acidity content in the northeastern United States was calculated by Cogbill and Likens[89] to be 62 percent sulfuric acid, 32 percent nitric acid, and 6 percent hydrochloric acid (HCl). Despite its low but significant percentage in precipitation, HCl is a strong acid whose sources and mechanisms of formation have not been completely identified. Although the specific processes of HCl formation are not clearly understood, researchers have attempted to provide models to show the transformation of chlorine compounds or chlorine gas into hydrochloric acid, which is eventually absorbed by moisture and precipitated.

Precursors of HCl acid precipitation in the atmosphere are generated from ground-level natural and anthropogenic sources. Table 2-11 provides a list of some of the major contributors of chlorides that may undergo reactions to form HCl.

Yue et al.[95] used a theoretical model showing that the formation of HCl is dependent on the interactions of SO_2, ammonia, CO_2, oxygen, and sulfuric acid. Cauer[96] suggested that salt spray oxidized by ozone and photochemically hydrolyzed into HCl is subsequently absorbed by moisture, resulting in acidic precipitation. However, Kohler and Bath[97] found that the sea salt (NaCl) to HCl conversion does not fully account for changes of the Na^+/Cl^- ratio in the air. Another model[94] proposed that particulate chloride interacts with NO_2 to produce HCl. Robbins, et al.[98] suggested that the reaction of aqueous HNO_3 with NaCl produces HCl in the atmosphere.

The natural sources of chloride include salt spray from the oceans[72,99] volcanic gases,[100] and upper atmospheric reactions.[97] The occurrence of chlorine gas in air is rare because of its highly reactive characteristics,[100,101] therefore, chloride compounds are more likely to be found. Concentrations of chlorides in coastal and noncoastal areas range from 0.02 mg/m^3 to 44 mg/m^3. Salt spray emissions estimated on a global scale are 600 to 1500 x 10^6 metric tons per year.[14,92,102] Nordlie[103] measured volcanic activity and found that chlorides are released in the magmatic gases. Bartel[93] estimated the global volcanic contribution of chloride to be 7.6 x 10^6 metric tons per year. Duce[94] estimated that 600 x 10^6 metric tons per year of HCl are produced through the interaction of trace gases and precipitation.

Anthropogenically produced chlorine and chlorides are emitted in various manufacturing and process operations; primarily in the manufacturing, handling, and liquefaction of chlorine gas and HCl. Annual production rates of chlorine and HCl in the United States have been increasing since 1978.[104,105] Chlorine is used in a variety of industries including the production of solvents, pesticides, herbicides, chlorinated hydrocarbons, plastics, and bleaches. Other uses of chlorine compounds are in wastewater treatment, in pulp and paper mills, and in fluxing of ferrous and nonferrous metals.

TABLE 2-11. EMISSION OF CHLORIDE PRECURSORS TO THE ATMOSPHERE

Source	U.S. Emission rate (10^3 metric tons)	Global emission rate (10^3 metric tons)
Anthropogenic sources		
Chlorine gas and liquid manufacture	42.7[a] as Cl_2	
HCl acid manufacture	0.73[a] as Cl_2	
Chlorinated hydrocarbon manufacture	206.6[b] as HCl	
Pulp and paper mills	16[a] as Cl_2	NA
Ferrous metal fluxing	0.09[a] as Cl_2	
Nonferrous metal fluxing	1.72[a] as Cl_2	
Coal combustion	NA	
Water treatment	NA	
Natural sources		
Ocean salt spray		600,000[c] – 1,500,000[d] as NaCl
Volcanic gases	NA	7,600[e,f]
Atmospheric reactions		600,000[g] as HCl

[a] Sittig (1975).[90]
[b] Khan and Hughes (1979).[91]
[c] Junge (1963).[14]
[d] Eriksson (1959).[92]
[e] Bartel (1972).[93]
[f] Chloride form not stated.
[g] Duce (1969).[94]
NA = Not available.

The chlorine production facilities are located mainly in the eastern United States with a concentration of plants in the Gulf Coast region of Texas, Louisiana, and Alabama, which accounted for nearly 77 percent of the total chlorine output in 1975.[104] Other facilities are distributed throughout the remainder of the eastern United States, on the West Coast, and in the Midwest. Hydrochloric acid production facilities are distributed in regions near chlorine production plants and are located in 29 states in the United States. Chlorinated hydrocarbon facilities are located mainly in the northeastern, south central, and western United States. Texas and Louisiana have the largest number of facilities on a per-state basis compared to the rest of the nation.

The combustion of coal by power generating facilities also releases chlorides into the atmosphere. The chloride released and its subsequent dissolution in moisture can yield hydrochloric acid in precipitation.[106] The Tennessee Valley Authority has reported the presence of HCl in flue gases from coal combustion up to concentrations of 50 to 75 ppm.[88] In urban Britain, the deposition of chlorides was attributed to the chlorine in coals (up to 1 percent by weight)[107] because the high levels measured in precipitation in this region could not be accounted for by marine sources alone.[108] In the United States, the release of chlorides from coal combustion has been noted, given that coals mined in central U.S. and Appalachia contain between 0.01 to 0.5 percent chlorine by weight.[109]

Synergistic Effects

Other pollutants that can potentially affect the formation and magnitude of acid precipitation and its precursors are emitted to and are present in the atmosphere. Although the potential synergistic reactions of these substances are largely a matter of speculation and hypothesis at this time, further research can be expected to reveal the extent of their contributions to the acid rain problem. Details on the atmospheric transformation and transport, scavenging, and removal of these pollutants and the precursors of acid rain are covered in Section 3.

Ozone--

Ozone and other photochemical oxidants may play a role in the conversion of SO_x and NO_x to sulfates and nitrates. Chemical reactions leading to the formation of these acid rain precursors have been associated with gas phase and heterogeneous (gas-liquid and gas-solid) processes related to photochemical smog. The major components of photochemical smog are NO_x and hydrocarbons, which are produced in large part by transportation sources in urban population centers.

Widespread regional exposure to ozone has been coupled with high concentrations of airborne sulfates. One study of summertime aerosol formation in urban plumes observed that the greatest rates of formation of $SO_4^=$ occurred when the plume was tracked through an ozone-laden air mass.[110] In addition, conversion rates of sulfur dioxide to sulfate have been observed to be higher in the daytime, suggesting a photochemical mechanism.[88,111-113]

Indirect photo-oxidation of SO_2 in the gas phase is a major route for conversion to $SO_4^=$. Direct collision of SO_2 with strong oxidizing radicals such as $HO\cdot$, $HO_2\cdot$, and $CH_3O_2\cdot$ results in the conversion. These radicals are intermediate products of the hydrocarbon-NO_x reactions that occur in photochemical smog. Oxidation rates produced by the presence of gas phase radicals may be as high as 10 percent per hour, depending on free radical concentrations.[114]

Oxidation of SO_2 by ozone in the liquid phase may also occur.[115,116] Ozone absorbed in liquid droplets can promote oxidation to rates that exceed the conversion by indirect photo-oxidation. The reaction rates are relatively high at low pH levels in the aqueous phase, but the variability of existing rate data prevents an accurate assessment of this mechanism in the atmosphere.[111] Another factor that may decrease the importance of this conversion mechanism is the scavenging of atmospheric ozone by NO in power plant plumes. Depression of ozone levels by this reaction may result in only minor enhancement of SO_2 oxidation.[117]

Although ozone is formed by the photochemical reactions of hydrocarbons and NO_x, it may also play a role in the conversion of NO_x to nitrates. The major removal pathway for atmospheric NO_2 is by further oxidation to nitric acid, which is readily dry deposited, and NO_3^- aerosols, which are removed by precipitation. Oxidation rates depend on atmospheric photochemistry and the availability of atmospheric ozone.[114] A complex series of reactions is involved, and NO_x may be switched back and forth between various oxidation states over periods ranging from hours to days.[118]

In addition to nitric acid, which is among the pollutants formed directly in the photochemical reaction mix, the scavenging of ozone by NO in plumes from major fuel burning installations may lead to further nitric acid production. Ozone reacts readily with NO to form NO_2 and molecular oxygen. Additional ozone may react with NO_2 to form a transient symmetrical nitrogen trioxide species, which then reacts with NO_2 and water to form nitric acid. This reaction is known to occur homogeneously (in the gas phase) and heterogeneously, although the latter may predominate.[118] Nitric acid also may be produced by the reaction of NO_2 with hydroxyl radicals, although free radical concentrations are expected to be lower in power plant plumes than in a photochemical reaction mix.

Carbon Dioxide--

Pure distilled water has a pH level of 7. In the atmosphere, various gases are absorbed and dissolved in moisture because of atmospheric pressure, resulting in a gas-liquid equilibrium. Carbon dioxide is soluble in water as either aqueous CO_2 or as carbonic acid, H_2CO_3, a weak acid. Using a simplified model of water in equilibrium with atmospheric CO_2 (0.032 percent by volume) at 25°C, the water will have a pH of 5.65.[119,120] However, this background or baseline pH level is a function not only of CO_2, but of other gases and soluble particles.[121] It can be expected that the extent to which SO_x, NO_x, ammonia and chlorides contribute to the acidification of rainwater will vary according to the background pH. This synergism apparently disappears below a pH of 5. An experiment conducted by Galloway, et al.[121] found that carbonic acid, which is in equilibrium with the atmosphere, had no influence on the measured pH of any aqueous samples tested below a pH of 5.

In addition, the presence of dissolved carbonate species and the exchange of CO_2 with the atmosphere allows certain bodies of water to have a buffer mechanism with capacities for both acid and base neutralization. However, for lakes and other bodies of water having very little dissolved carbonates, the presence of carbon dioxide tends to acidify the water to the 5.65 level.[119] This is discussed further in Section 5, Mitigative Strategies.

Particulates--
Another factor in the formation or neutralization of acid precipitation is the presence of natural and anthropogenic dust. It has been suggested that alkaline dusts may react with and neutralize strong acids in the atmosphere.[122] Natural sources of atmospheric dust include soil erosion, agricultural activities, construction activities, volcanoes, and cosmic material.[122] Terrestrial dusts from the first three sources are composed of quartz, feldspar, carbonate minerals, dolomite, calcite, and clay. In agricultural areas of the central and midwestern United States, dust emissions from calcite and carbonate deposits act as buffers.[123] Natural dusts in the southwestern and central plains regions are characterized by higher emissions and higher alkalinity (pH 7 to 8), whereas eastern dusts are slightly acidic (pH 6 to 7).[123] In areas of dense forest cover, only small quantities of dust are generated and the soil is non-alkaline.

Adomaitis[124] investigated the relationship between pH and the chemical composition of snow. Dusty snows, characterized by high soluble mineral content, were found to be alkaline, whereas clean snows with a lower soluble cation content were slightly acidic (pH of 5).

Shannon and Fine[125] have measured pH values of 9 to 11 from water extracts of fly ash. Lower concentrations of Na_2O and CaO in eastern coal fly ashes result in less alkaline ash from eastern coals.[126] Likens and Borman have suggested that the removal of alkaline fly ash emissions with industrial electrostatic precipitators contributes to acid rain formation.[127] On the other hand, Ananth, et al.[126] have indicated that the removal of fly ash may not be a contributor because the particles removed by the precipitator are probably larger in size than those remaining in the plume and, thus, if left uncontrolled, would settle more rapidly, viz, reduced residence time in the plume. They further indicate that there is insufficient data to determine whether the alkalinity of fly ash has a significant effect on the neutralization of acid rain.

Fly ash may also enhance the formation of acid rain. Ananth et al.[126] investigated two mechanisms by which particulate matter from coal combustion may enhance the oxidation of SO_2. Absorption-oxidation of SO_2 involves catalysis by suspended fly ash in the presence of large amounts of water. Such conditions may occur in the liquid phase, in cloud droplets, or during periods of actual precipitation. A second mechanism occurs in the presence of small quantities of water. This mechanism, direct catalytic oxidation, may rely on metallic constitutents of the fly ash acting as catalysts.[118] Although the catalytic activity of many trace and minor elements has been demonstrated, the activity of vanadium pentoxide is considered to be much

higher than that of other metals.[126] Order of magnitude calculations by
Ananth, et al.[126] indicate that both absorption-oxidation and direct catalytic
oxidation are plausible mechanisms for rapid formation of $SO_4^=$ in power plant
plumes. Vanadium pentoxide is also formed in the combustion of residual oil
and, therefore, may influence the fate of SO_2 from oil-fired power plants. The
study of the catalytic oxidation of SO_2 to $SO_4^=$ needs much greater study, as it
may be an important route in the transformation of SO_2 to $SO_4^=$ particulates.

Atmospheric reactions that lead to the conversion of SO_x and NO_x to sulfate and nitrate aerosols have important implications. The conversion of acid rain precursors to these more stable particulate forms increases their atmospheric lifetime, facilitates transport, and contributes to the regional nature of the acid rain problem.[128]

SUMMARY

The principal contributors to precipitation acidity are sulfuric and
nitric acid. Although no direct quantitative relationship between pollutant
emissions and the measured acidity of precipitation has been demonstrated,
naturally and anthropogenically produced SO_x and NO_x are considered major acid
rain precursors. The most recent estimates indicate that two-thirds of global
SO_x emissions are manmade, although in highly industrialized areas, anthropogenic sources may account for 90 percent of the total. Coal and oil combustion
and various industrial processes are the major contributors. Estimates of the
natural source contribution to global NO_x emissions vary considerably, although
fossil fuel combustion and transportation sources account for most of the manmade emissions.

Emission trends for SO_x and NO_x in the United States shows decreases for
both of these pollutants during the 1970's. Although it appears that transportation will continue to be a major source of NO_x emissions, future SO_x and NO_x
emissions from fuel combustion will be influenced by a number of factors. The
relative amounts of coal, oil, and natural gas that will be burned as well as
the proportions of old and new boilers are important. Coal cleaning, the use
of low sulfur coal, the application of control technologies such as flue gas
desulfurization systems, and the application of new combustion technologies,
such as fluidized-bed combustion, will all play a role in determining fuel
combustion SO_x and NO_x emissions.

In addition to SO_x and NO_x, other pollutants may directly or indirectly
affect the acidity of precipitation. Hydrochloric acid, produced from atmospheric chlorine and chlorides, accounts for a small percentage of rainwater
acidity. Ammonia, because of its alkaline nature, is able to neutralize acidic
components of precipitation. Less well understood are the roles of ozone,
carbon dioxide, and particulate matter. Alkaline natural dusts and fly ash
emissions may be capable of neutralizing precipitation acidity. However, certain metallic constituents of fly ash, such as vanadium pentoxide from residual
oil, may be involved in the catalytic oxidation of SO_2 to sulfates, leading to
a possible increase in acidity.

A number of pollutants that may be involved in the formation or neutralization of acid rain have been identified. However, a number of uncertainties remain. Precise quantification of the contribution of each pollutant, the relative importance of their various sources, and determination of the factors affecting future emissions are all subjects for further investigation.

One study indicates that the pH of precipitation is an integrated result of both the natural and man-made contributions to acidity.[114] It states that "natural precipitation" without the addition of alkaline materials should have a pH from 4.5 - 5.5 (this is the equilibrium value of NO_2 and water vapor; the corresponding value for SO_2 is 4.1 - 5.1).[115]

REFERENCES

1. U.S. Environmental Protection Agency. 1977 National Emissions Report. In: National Emissions Data System of the Aerometric and Emissions Reporting System. EPA-450/4-80-005, 1980.

2. Galloway, J. N., and D. M. Whelpdale. An Atmospheric Sulfur Budget for Eastern North America. Atmospheric Environment, 14:409-417, 1980.

3. Granat, L., H. Rodhe, and R. O. Hallberg. The Global Sulfur Cycle. In: Nitrogen, Phosphorus, and Sulfur - Global Cycles, Svensson and Söderlund, eds. SCOPE Report 7, Ecol. Bull. (Stockholm), 22:89-134, 1976.

4. Kellogg, W. D., R. D. Cadle, E. R. Allen, A. L. Lazrus, and E. A. Martell. The Sulfur Cycle. Science, 175:587-596, 1972.

5. Hitchcock, D. R. Dimethyl Sulfide Emissions to the Global Atmosphere. Chemosphere, 3:137-138, 1975.

6. Adams, D. F., M. R. Pack, W. L. Bamesberger, A. E. Sherrard, and S. O. Farwell. Measurement of Biogenic Sulfur-Containing Gas Emissions from Soils and Vegetation. In: 71st Annual Meeting of the Air Pollution Control Assoc., Houston, Texas, Paper No. 78-7.6, 1978.

7. Maroulis, P. J., and A. R. Bandy. Estimate of the Contribution of Biologically Produced Dimethyl Sulfate to the Global Sulfur Cycle. Science, 196:647-648, 1977.

8. Jaescheke, W., and W. Haunold. New Methods and First Results of Measuring Atmospheric H_2S and SO_2 in the ppb Range. In: Special Environmental Report No. 10 - Air Pollution Measurement Techniques, WMO No. 460, 1977. pp. 193-198.

9. Hansen, M. H., K. Ingvorsen, and B. B. Jorgensen. Mechanisms of Hydrogen Sulfide Release from Coastal Marine Sediments to the Atmosphere. Limno. Oceanogr., 23:68-76, 1978.

10. Liss, P. S., and P. G. Slater. Flux of Gases Across the Air-Sea Interface. Nature (London), 247:181-184, 1974.

11. Friend, J. P. The Global Sulfur Cycle. In: Chemistry of the Lower Atmosphere, S. I. Rasool, ed. Plenum Press, New York, 1973. pp. 177-201.

12. Aneja, V. P., J. H. Overton, L. T. Cupitt, J. L. Durham, and W. E. Wilson. Direct Measurements of Emission Rates of Some Atmospheric Biogenic Compounds. In: 174th American Chemical Society Meeting, Miami, Florida, 1978.

13. Eriksson, E. The Yearly Circulation of Chloride and Sulfur in Nature; Meteorological, Geochemical, and Pedological Implications. 2, Tellus, 12:63, 1960.

14. Junge, C. E. Atmospheric Chemistry and Radioactivity. Academic Press, New York, New York, 1963.

15. Robinson, E., and R. C. Robbins. Sources, Abundance, and Fate of Gaseous Atmospheric Pollutants. Stanford Research Institute Final Proj. Rep PR-6755, Menlo Park, California, 1968.

16. Robinson, E., and R. C. Robbins. Emissions, Concentrations, and Fate of Gaseous Atmospheric Pollutants. In: Air Pollution Control, Part II, W. Strauss, ed. Wiley-Interscience, New York, New York, 1972. pp. 1-93.

17. Whelpdale, D. M. Atmospheric Pathways of Sulfur Compounds. In: MARC Report No. 7, Monitoring and Assessment Research Centre, University of London, London, United Kingdom, 1978.

18. Robinson, E., R. B. Husar, and J. N. Galloway. Sulfur Oxides in the Atmosphere. In: Sulfur Oxides, National Academy of Sciences, Washington, D.C., 1978.

19. Robinson, E., and R. C. Robbins. Gaseous Atmospheric Pollution from Urban and Natural Sources. In: The Changing Global Environment, S. F. Singer, ed. D. Reidel Publ. Co., Dardrecht-Holland, Boston, 1975. pp. 111-123.

20. Söderlund, R., and B. H. Svensson. The Global Nitrogen Cycle. In: Nitrogen, Phosphorus, and Sulfur - Global Cycles. SCOPE Report 7, B. H. Svensson and R. Söderlund, eds. Ecol. Bull. (Stockholm), 22:23-73. 1976.

21. Delwiche, C. C. The Nitrogen Cycle. Scientific American, 223(3):137-146, 1970.

22. Burn, R. C., and R. W. F. Hardy. Nitrogen Fixation in Bacteria and Higher Plants. Springer-Verlag, Berlin-Heidelberg-New York, 1975.

23. Liu, S. C., R. J. Cicerone, R. M. Donahue, and W. L. Chameides. Sources and Sinks of Atmospheric N_2O and the Possible Ozone Reduction due to Industrial Fixed Nitrogen Fertilizers. Tellus, 29:251-263, 1977.

24. Chameides, W. L., D. H. Stedman, R. R. Dickerson, D. W. Rusch, and R. J. Cicerone. NO_x Production in Lighting. J. Atmos. Sci., 34:143-149, 1977.

25. Crutzen, P. J., and D. H. Ehhalt. Effects of Nitrogen Fertilizers and Combustion in the Stratospheric Ozone Layer. Ambio., 6(2-3):112-117, 1977.

26. Noxon, J. F. Atmospheric Nitrogen Fixation by Lightning. Geophys. Res. Lett., 3:463-465, 1976.

27. Griffing, G. W. Ozone and Oxides of Nitrogen Production During Thunderstorms. J. Geophy. Res., 82:943-950. 1977.

28. Noxon, J. F. Tropospheric NO_2. J. Geophys. Res., 83:3051-3057, 1978.

29. Crutzen, P. F., I. S. A. Isaksen, and J. R. McAfee. The Impact of the Chlorocarbon Industry on the Ozone Layer. J. Geophys. Res., 83:345, 1978.

30. U.S. Environmental Protection Agency. National Air Quality, Monitoring and Emissions Trends Report, 1977. EPA-450/2-78-052, 1978.

31. Department of Health, Education, and Welfare National Emission Standards Study. Appendix - Volume 3, Appendix F (Part 2), 1970.

32. U.S. Environmental Protection Agency. National Air Pollutant Emission Estimates, 1940-1976. EPA-450/1-78-003, 1978.

33. Hendrey, G. R., and F. W. Lipfert. Acid Precipitation and the Aquatic Environment. Brookhaven National Laboratory. Presented to the Committee on Energy and Natural Resources, United States Senate, May 28, 1980.

34. The Regional Issue and Assessment (RIA) Program, Second Annual Report. Prepared for the U.S. Department of Energy. Draft Report. 1980.

35. Lin, K., and J. Dotter. Steam Electric Plant Factors 1979. National Coal Association, 1979.

36. Voldner, E. C., Y. Shah, and D. M. Whelpdale. A Preliminary Canadian Emissions Inventory for Sulfur and Nitrogen Oxides. Atmos. Environ., 14(4):419-428, 1980.

37. Environment Canada. A Nation-wide Inventory of Air Contaminant Emissions - 1974. Report EPS 3-AP-78-2, Air Pollution Control Directorate, 1978. Augmented and updated by P. J. Choquette, 1979.

38. Altshuller, A. P., and G. A. McBean. The LRTAP Problem in North America: A Preliminary Overview. Prepared by the United States - Canada Research Consultation Group on the Long-Range Transport of Air Pollutants.

39. Battelle Memorial Institute. The Federal R&D Plan for Air Pollution Control by Combustion Processes Modification. Contract No. CPA-22-69-147, 1971.

40. Locklin, D. W., et al. An Overview of Research Needs for Air Pollution Control by Combustion Process Modification. AICHE Symp. Series, 68(126):1, 1972.

41. Commission on Natural Resources. Air Quality and Stationary Source Emission Control. National Academy of Sciences, 1975.

42. Demeter, J., and D. Bienstock. Sulfur Retention in Anthracite Ash. Report of Investigation 7160, U.S. Bureau of Mines, 1968.

43. Sondreal, E. A., W. R. Kube, and J. L. Elder. Analysis of the Northern Great Plains Province Lignites and Their Ash: A Study of Variability. U.S. Bureau of Mines Report of Investigations 7158, 1968.

44. Ode, W. H., and F. H. Gibson. Effect of Sulfur Retention on Determined Ash in Lower-Rank Coals. U.S. Bureau of Mines of Investigations 5931, 1962.

45. Walden Research Corporation. Final Report on Sulfur, Mercury, and Other Materials Studies at Neil Simpson Station. Cambridge, Massachusetts, 1973.

46. Gronhovd, G. H., P. H. Tufte, and S. J. Selle. Some Studies on Stack Emissions from Lignite-Fired Powerplants. In: 1973 Lignite Symposium, Grand Forks, North Dakota, 1973.

47. Homolya, J. B., and J. L. Cheney. An Assessment of Sulfuric Acid and Sulfate Emissions from the Combustion of Fossil Fuels. In: Workshop Proceedings on Primary Sulfate Emissions from Combustion Sources, Volume 2, Characterization, EPA-600/9-78-020b, U.S. Environmental Protection Agency, 1978. pp. 3-11.

48. Bennett, L., and K. T. Knapp. Sulfur and Trace Metal Particulate Emissions from Combustion Sources. In: Workshop Proceedings on Primary Sulfate Emissions from Combustion Sources, Volume 2, Characterization. EPA-600/9-78-020b, U.S. Environmental Protection Agency, 1978. pp. 165-183.

49. Homolya, J. B., H. M. Barnes, and C. R. Fortune. A Characterization of the Gaseous Sulfur Emissions from Coal and Oil-Fired Boilers. In: Energy and the Environment, Proceedings of the 4th National Conference, October 3-7, 1976.

50. Bueters, K. A., and W. W. Habect. NO_x Emissions from Tangentially Fired Utility Boilers-Theory. In: 66th Annual AICHE Meeting. Philadelphia, Pennsylvania, 1973.

51. Crawford, A. R., E. H. Manny, and W. Bartok. Field Testing: Application of Combustion Modifications to Control NO_x Emissions from Utility Boilers. Exxon Research and Engineering Company, 1974.

52. Compilation of Air Pollutant Emission Factors (with supplements). EPA Publication AP-42, U.S. Environmental Protection Agency, 1975.

53. Surprenant, N., R. Hall, S. Slater, T. Susa, M. Sussman, and C. Young. Preliminary Emissions Assessment of Conventional Stationary Combustion Systems: Vol. II. EPA-600/2-76-046b, U.S. Environmental Protection Agency, 1976.

54. Air Quality and Stationary Source Emission Control. Report by Commission on Natural Resources. National Academy of Sciences, March 1975.

55. Blakeslee, C. E., and H. E. Burbach. NO_x Emissions from Tangentially-Fired Utility Boilers - Practice. Presented at the 66th Annual AICHE Meeting, Philadelphia, Pa. November 1973.

56. Dickerman, J. C., and K. L. Johnson. Technology Assessment Report for Industrial Boiler Applications: Flue Gas Desulfurization. EPA-600/7-79-178i, U.S. Environmental Protection Agency. 1979.

57. McCurley, W. R., and D. G. DeAngelis. Measurement of Sulfur Oxides from Coal-Fired Utility and Industrial Boilers. In: Workshop Proceedings on Primary Sulfate Emissions from Combustion Sources, Vol. 2, Characterization, EPA-600/9-78-020b, U.S. Environmental Protection Agency, 1978. pp. 67-85.

58. Homolya, J. B., and J. C. Cheney. A Study of Primary Sulfate Emissions from a Coal-Fired Boiler with PCB. J. Air. Pollut. Control Assoc., 29, 1979.

59. U.S. Department of Energy. Energy Data Reports. EIA-0049/1, 1976.

60. California Board of Air Sanitation. The Oxides of Nitrogen in Air Pollution. California Department of Public Health, 1966.

61. Ashby, H. A., R. C. Stahman, B. H. Eccleston, and R. W. Hurn. Vehicle Emissions - Summer to Winter. SAE Paper 741053, SAE Automobile Engineering Meeting, Toronto, Canada, 1974.

62. Federal Highway Administration. Traffic Volume Trends, Tables 5A, 5B, and 9A. November-December 1978. Highway Statistics Division, Washington, D.C.

63. Lavery, T. F., G. M. Hidy, R. L. Baskett, and P. K. Mueller. The Formation and Regional Accumulation of Sulfate Concentrations in the Northeastern United States. Prepared for the Proceedings of the Symposium on Environmental and Climatic Impact of Coal Utilization. 1979.

64. U.S. Environmental Protection Agency. Background Information for New Sources Performance Standards: Primary Copper, Zinc, and Lead Smelters, Vol. I, Proposed Standard. EPA-450/2-74-002a, U.S. Environmental Protection Agency, 1974.

65. Klemm, H. A., and R. J. Brennan. Emission Inventory on the SURE Region. GCA/Technology Division, EPRI Contract TP 862-5, 1980.

66. Dykema, O. W., and V. E. Kemp. Inventory of Combustion-Related Emissions from Stationary Sources (First Update). EPA-600/2-77-066a, U.S. Environmental Protection Agency, 1977.

67. Clark, T. L. Gridded Annual Air Pollutant Emissions East of the Rocky Mountains. EPA-600/4-79-030, U.S. Environmental Protection Agency, 1979.

68. GCA/Technology Division. Unpublished data from the development of the SURE Emissions Inventory.

69. Environmenal Research & Technology, Inc. The Sulfate Regional Experiment (SURE) Final Report, Volume 2. Project No. EPRI RP 862, 1980.

70. Liljestrand, H. M., and J. J. Morgan. Chemical Composition of Acid Precipitation in Pasadena, CA. Environ. Sci. & Tech., 12 (12), 1978.

71. Liptak, B. G., ed. Environmental Engineer's Handbook, Vol. 2, Air Pollution. Chilton Book Co., Radnor, Pennsylvania, 1974.

72. Likens, G. E., R. F. Wright, J. N. Galloway, and T. J. Butler. Acid Rain. Scientific American, 241(4), 1979.

73. Scott and Lamb. J. of Am. Chem. Soc., 92-3943, 1970.

74. National Center for Atmos. Res., unpublished results. (See Reference 66).

75. Doyle, G. J., et al. Simultaneous Concentrations of NH_3 and HNO_3 in a Polluted Atmosphere and Their Equilibrium Relationship to Particulate Ammonium Nitrate. Env. Sci. & Tech., 13:1416, 1979.

76. McConnell, J.C. Atmosphere Ammonia. J. Geophys. Res., 78:7812-7820, 1973.

77. Stanford, G., and J. O. Legg, et al. Denitrification and Associated Nitrogen Transformation in Soils. Soil Sci., 120:147-152, 1975.

78. Denmead, O. T., J. R. Simpson, and J. R. Freney. Ammonia Flux into the Atmosphere from Grazed Pasture. Science, 185:609-610, 1974.

79. Bouldin, D. R., et al. Losses of Inorganic Nitrogen from Aquatic Systems. J. Env. Qual., 3(2):107-114, 1974.

80. Pelczar, M. J., Jr., R. D. Reid, and E. C. S. Chan. Microbiology. Fourth Edition. McGraw-Hill Book Company, New York, New York, 1977.

81. Wark, K., and C. F. Warner. Air Pollution, Its Origin and Control. Harper and Row Publishers, Inc., New York, New York, 1976.

82. Health Effects Research Laboratory. Ammonia. U.S. Environmental Protection Agency, EPA-600/1-77-054, 1977.

83. National Research Council. Ammonia. Subcommittee on Ammonia, Committee on Medical and Biological Effects of Environmental Pollutants, Assembly of Sciences, National Academy of Science, 1978.

84. Rawlings, G. D., and R. B. Reznik. Source Assessment: Synthetic Ammonia Production. Monsanto Research Corp., IERL, EPA-600/2-77-107m, U.S. Environmental Protection Agency, 1977.

85. Council for Agricultural Science and Technology. Effect of Increased Nitrogen Fixation on Stratospheric Ozone. In: CAST Report No. 53, 1976.

86. National Research Council. Nitrates: An Environmental Assessment. National Academy of Sciences, Washington, D.C., 1978.

87. Frank, C. R., and G. K. Voight. Potential Effects of Acid Precipitation on Soils in the Humid Temperature Zone. In: Proceedings of the First International Symposium on Acid Precipitation and the Forest Ecosystem. Columbus, Ohio. May 12-15, 1975.

88. Gorham, E. Acid Precipitation and Its Influence Upon Aquatic Ecosystems-- An Overview. In: Proc. First International Symposium on Acid Precipitation and the Forest Ecosystem, L. S. Dochinger and T. A. Selinga, eds. Columbus, Ohio, 1975.

89. Cogbill, C. V., and G. E. Likens. Acid Precipitation in the Northeastern United States. Water Resources Res., 10(6):1133-7, 1974.

90. Sittig, M. Environmental Sources and Emissions Handbook. Noyes Data Corp., Park Ridge, New Jersey, 1975.

91. Kahn, Z. S., and T. W. Hughes. Source Assessment: Chlorinated Hydrocarbons Manufacture. Monsanto Research Corp. EPA-600/2-79-019. U.S. Environmental Protection Agency, 1979.

92. Eriksson, E. The Yearly Circulation of Chloride and Sulfur in Nature; Meteorological, Geochemical and Pedological Implications, Part I. Tellus, 11:375, 1959.

93. Bartel, O. G. Health Phys., 22:387, 1972.

94. Duce, R. A. On the Source of Gaseous Chlorine in the Marine Atmosphere, J. Geophys. Res., 74:4597-9599, 1969.

95. Yue, G. K., V. A. Mohen, and C. S. Kiang. A Mechanism for Hydrochloric Acid Production in Clouds. In: Proc. of the First International Symposium on Acid Precipitation and the Forest Ecosystem, Columbus, Ohio, May 12-15, 1975.

96. Cauer, H. Einiges uber dan Einpluss des Meeres aur den Chemisus der Luft, der Balneologes, 5:409-415, 1938.

97. Kohler, H., and M. Bath. Qualitative Chemical Analysis of Condensation Nuclei from Sea Water. Nova Acta Regiol. Soc. Scient., 15(7):24, 1953.

98. Robbins, R. C., R. D. Cadle, and D. L. Eckhart. The Conversion of NaCl to Hydrogen Chloride in the Atmosphere. J. of Meteorology, 16(53), 1959.

99. Laird, A. R., and R. W. Miksad. Observations on the Particulate Chlorine Distribution in the Houston-Galveston Area. Atmos. Envir. 12:1537-1542, 1978.

100. Stah., Q. L., and Litton Systems, Inc. Preliminary Air Pollution Survey of Chlorine Gas. APTD 69-33, U.S. Department of Health, Education, and Welfare, 1969.

101. Barton, L. V. Toxicity of Ammonia, Chlorine, Hydrogen Cyanide, Hydrogen Sulfide, and Sulfur Dioxide Gases. Contrib. Boyce Thompson Inst., 11(5):357, 1940.

102. Seinfeld, J. H. Air Pollution, Physical and Chemical Fundamentals. McGraw-Hill, Inc., New York, New York, 1975.

103. Nordlie, B. E. Amer. J. Sci., 271-417, 1971.

104. Kirk and Othmer. Encyclopedia of Chemical Technology, 3rd Ed., 1(800) J. Wiley & Sons, New York, New York, 1978.

105. USDC. Survey of Current Business, 60(6), 1980.

106. Babich, H., D. L. Davis, and G. Stotzky. Acid Precipitation. Envir., 24(4):6, 1980.

107. Meetham, A. R. Natural Removal of Pollution from the Atmosphere. Quart. J. Roy. Metor. Soc., 76:359-371, 1950.

108. Parker, A. The Investigation of Atmospheric Pollution, 27th Report of the Standing Conference of Cooperating Bodies. HMSO, London, 1955.

109. Coal Research Center. Pollution by Chlorine in Coal Combustion. In: Sect. V of Air Pollution Research Progress Report for Quarter Ended December 31, 1966. BM/41-BM/50, U.S. Bureau of Mines, 1966.

110. Loiy, P. J., and T. J. Kneip. Aerosols: Antrhopogenic and Natural Sources and Transport - Summary of a Conference. J. Air Pollution Control Assoc., 30:358-361, 1980.

111. Committee on Sulfur Oxides. In: Sulfur Oxides, National Academy of Sciences, Washington, D.C., 1978.

112. Roberts, D. B., and D. J. Williams. The Kinetics of Oxidation of Sulphur Dioxide Within the Plume from a Sulphide Smelter in a Remote Region. Atmos. Environ., 13:1485, 1979.

113. Hegg, D. A., and P. V. Hobbs. Measurement of Gas-to-Particle Conversion in the Plumes from Five Coal-Fired Electric Power Plants. Atmos. Environ., 14:99, 1980.

114. Barnes, R. A. The Long Range Transport of Air Pollution: A Review of European Experience. J. Air Pollution Control Assoc., 29:1219-1235, 1979.

115. Beilke, S., and G. Gravenhorst. Heterogeneous SO_2-Oxidation in the Droplet Phase. Atmos. Environ., 12:231-239, 1978.

116. Eggleton, A. E. J., and R. A. Cox. Homogeneous Oxidation of Sulphur Compounds in the Atmosphere. Atmos. Environ, 12:227-230, 1978.

117. Drewes, D. R., J. M. Hales, and C. Hakkarinen. SO_2 Oxidation in Precipitation Falling Through a Power Plant Plume. In: the Commission on Atmospheric Chemistry and Global Pollution Symposium on the Budget and Cycles of Trace Gases and Aerosols in The Atmosphere, University of Colorado, Boulder, Colorado, 1979.

118. Haagen-Smit, A. J., and L. G. Wayne. Atmospheric Reactions and Scavenging Processes. In: Air Pollution (Third Ed.), Volume I, A. C. Stern, ed. Academic Press, New York, New York, 1976. Ch. 6

119. Stumm, W., and J. J. Morgan. Aquatic Chemistry. Wiley Interscience, New York, New York, 1970.

120. Glass, N. R., G. E. Glass, and P. J. Rennie. Effects of Acid Precipitation. Environ. Sci. & Tech., 13(11), 1979.

121. Galloway, J. N., G. E. Likens, and E. S. Edgerton. Hydrogen Ion Speciation in the Acid Precipitation of the Northeastern United States. In: Proceedings of the First International Symposium on Acid Precipitation and the Forest Ecosystem. USDA Forest Service, Ohio State University, Columbus, Ohio, 1975.

122. Winkler, E. M. Natural Dust and Acid Rain. In: Proceedings of the First International Symposium on Acid Precipitation and the Forest Ecosystem, USDA Forest Service, Ohio State University, Columbus, Ohio, 1975.

123. Smith, R. M., P. C. Twiss, R. K. Krauss, and M. J. Brown. Dust Deposition in Relation to Site, Season, and Climatic Variables. Proc. Sci. Soc. Am., 34(11), 1970.

124. Adomaitis, V. A., H. A. Kantrud, and T. A. Shoesmith. Some Chemical Characteristics of Aeolian Deposits of Snow-Soil on Prairie Wetlands. Proc. North Dakota Acad. Sci., 21:65-69, 1967.

125. Shannon, D. G., and L. O. Fine. Cation Solubilities of Lignite Flyash. Envir. Sci. & Tech., 8:1026-1028.

126. Ananth, K. P., J. B. Galeski, L. J. Shannon, F. I. Honea, and D.C. Drehmel. Impact of Particulates and SO_2 from Coal-Fired Power Plants on Acid Rain. In: Proc. of the Fourth International Clean Air Congress, the Japanese Union of Air Pollution Prevention Association, 1977.

127. Likens, G. E., and F. H. Bormann. Acid Rain: A Serious Regional Environmental Problem. Science, 184:1176-1179.

128. Hidy, G. M., J. R. Mahoney, and B. J. Goldsmith. International Aspects of the Long Range Transport of Air Pollutants. Document P-5252, U.S. Department of State, 1978.

129. American Petroleum Institute. A Detailed Analysis of the Scientific Evidence Concerning Acid Precipitation, Comments on the Revised Air Quality Criteria For Oxides of Nitrogen, Appendix D, November 1980.

130. Lerman, A. Geochemical Processes: Water and Sediment Environments. Wiley-Interscience: New York 1979.

131. U.S. Department of Energy. Acid Rain: The Impact of Local Sources. Morgantown Energy Technology Center, Under Contract Number DE-AC-21-81MC-14787, November 1980.

Section 3

Atmospheric Transport, Transformation and Deposition Processes

Manmade pollutants are injected into the atmosphere at heights ranging from a few feet as is the case with auto exhaust, to more than 1000 feet in the case of tall stacks. Significant quantities of pollutants are also emitted from natural sources (see Section 2). The fate of all such pollutants depends on physical processes of dispersion, transport, and deposition and on complex chemical transformations that take place between source and final receptor. Their residence time within the atmosphere may be brief, as when emission takes place directly into an existing rainstorm, or may extend over several days or even weeks. In the latter case, the impact of pollutants, at least partially transformed, may reach distances hundreds or thousands of miles from the source. Methods to determine the relative contributions of possible sources to measured acidity are in early, exploratory stages. However, attempts are being made to describe the entire process using regional models. Some idea of the complexity of this undertaking, plus early modeling results, are provided in this section. This section also summarizes current interpretations of some of the monitoring results.

METEOROLOGICAL VARIABILITY

The atmosphere is a dynamic, constantly adjusting system, as evidenced by the passage of storms, ever changing wind fields, and changes in the vertical temperature structure brought about by the daily cycle of solar radiation. As a result, the simulation of the life cycle of pollutants from the time of their emission to the time of their deposition as acid precipitation by a storm system that may have developed several days later is an exceedingly complex task. In selecting meteorological inputs for use in regional models over such a time frame, major simplifications in the horizontal and vertical structure of the atmosphere are required. These simplifications must treat such features as wind shear, turbulence, convection, thermal layering, and boundary conditions. How this is done is alluded to briefly in the discussion of models presented later. Here, the discussion is limited to large-scale, climatalogical features of meteorological variability that play a part in the distribution of acid precipitation. Features briefly described here include prevailing wind patterns, storm tracks and the distribution of precipitation, and the occurrence of stagnant conditions.

Seasonal Changes in Mean Flow

Except for southern Florida and the Gulf Coast, the contiguous United States lies in the belt of the prevailing westerlies. Thus, on the average, unpolluted air enters the continent from the west and receives pollutants while traversing toward the east coast. However, this flow is impeded by mountain ranges, has day-to-day weather patterns superimposed, and undergoes significant seasonal change.

Seasonal differences are illustrated in Figures 3-1 and 3-2, which show the mean resultant surface winds at principal reporting stations during January and July. Large open arrows have been added to these maps to depict average flow conditions during these months of maximum seasonal difference. During both seasons, air enters the west coast from the Pacific; flow through the mountain states, although rather ill-defined because of topographical influences, shows an average movement from the south and west. Very marked seasonal differences occur east of the Rockies, however. During January, major inflow from Canada dominates the pattern over the north central and northeastern states. In the far south, flow from the Gulf of Mexico moves north over Texas, and air from the Atlantic crosses southern Florida.

In July, in the mean, all of the states east of the Rockies are dominated by southerly flow, which is particularly steady and strong over the central states and becomes southwesterly over the northeastern states. During the summer, incursions of Canadian air are less frequent than in winter and penetrate less deeply into the United States. Figure 3-3 represents surface flow over the North American continent schematically for the month of July, when sulfate deposition tends to be at a seasonal peak in the northeastern United States and eastern Canada.

Storm Tracks and Precipitation Patterns

Some of the most frequent routes of cyclonic weather systems are sketched in Figure 3-4. These paths give a general picture of the movement of low pressure centers and do not refer to the exact movement of individual storms. This figure shows that, regardless of their origin, many storms pass over the eastern states. In the development of storms moving from the west, moisture is supplied principally from the Gulf of Mexico or Atlantic Ocean, and the combination of this moisture supply and preferred storm paths leads to high annual precipitation amounts in the southern and eastern states, as shown in Figure 3-5. The second area with high annual amounts of precipitation lies along the coastal region of Washington and Oregon where storm activity from the Pacific is frequent, particularly during the winter and early spring. Figure 3-5 was prepared from mean precipitation amounts given for state climatic divisions, and precipitation amounts at specific locations may differ substantially from those indicated by the figure. The drying effect of the western mountain ranges on air masses moving in from the Pacific is reflected in the extensive areas in the west with less than 20 inches of annual precipitation.

The effectiveness of precipitation in cleansing the atmosphere depends not only on the total amount of precipitation that falls during a year, but also on its temporal distribution. The most readily available climatological statistic bearing on this distribution is the number of days in a year with measurable precipitation (i.e., $\geq$ 0.01 inches). Figure 3-6 gives the

Atmospheric Transport, Transformation and Deposition Processes   89

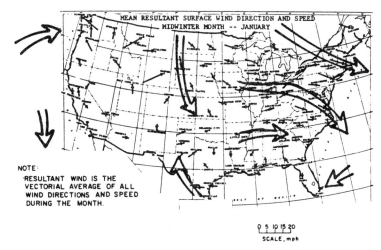

Figure 3-1. Mean resultant surface winds for January (Resultant wind map taken from Climatological Atlas of the United States[1]).

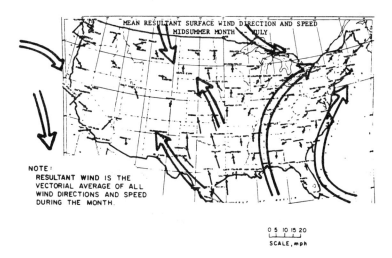

Figure 3-2. Mean resultant surface winds for July (Resultant wind map taken from Climatological Atlas of the United States[1]).

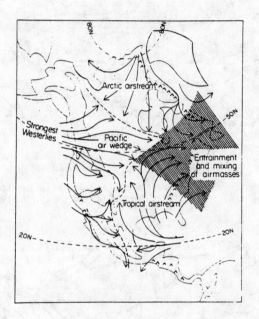

Figure 3-3. Schematic representation of the surface flow across North America, illustrating the wedge of Pacific air east of the Cordillera and the downstream entrainment and mixing of airstreams. Based on July resultant surface winds (Bryson et al.[2])

Atmospheric Transport, Transformation and Deposition Processes   91

Figure 3-4.   Frequent routes of storm centers.

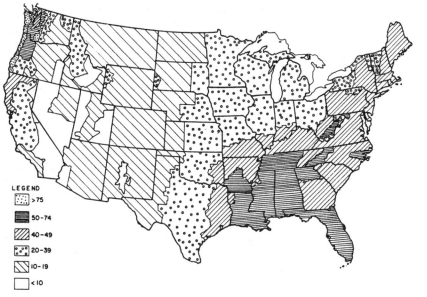

Figure 3-5.   Mean annual precipitation in inches (Prepared from data in Climatic Atlas of the United States[1]).

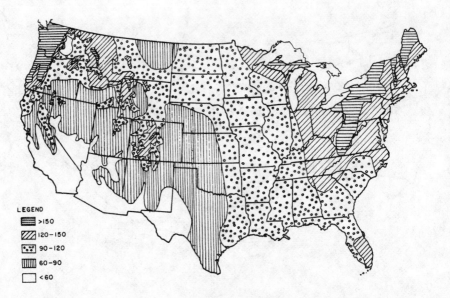

Figure 3-6. Mean number of days with 0.01 inches, or more of precipitation in 1 year (Prepared from data in Climatic Atlas of the United States[1]).

geographic distribution of this information. The number of days with precipitation and the annual precipitation have similar geographic patterns; the principal difference is in the eastern part of the country where total precipitation amounts are greatest in the south and the number of days with precipitation is greatest in the north. Thus the northeastern United States, which both produces acid rain precursor pollutants and is subject to an influx of acidic pollutants from neighboring regions, experiences frequent precipitation.

Occurrence of Stagnant Conditions

Warm high-pressure systems that are slow moving or stagnant favor the accumulation of sulfate compounds and other pollutants. The accompanying clear skies favor smog-producing photochemical reactions, and temperature inversions in the lower atmospheric layers limit the depth of vertical mixing. In addition, because of the absence of rain, depletion of the pollutant cloud is minimal.

Korshover[3] studied the frequency and persistence of stagnating anticyclones east of the Rockes during a 30-year period. He found that the number of cases per year when a stagnating anticyclone lasted for four or more days ranged from 4 to 16 and that the maximum frequency in both number of cases and number of days stagnation occurred in an area covering parts of Georgia, South Carolina, and North Carolina. Seasonally, the frequency of occurrence peaks in October and has a secondary maximum in May. One of the consequences of having an anticyclone with its clockwise circulation and light winds stagnate over the southeastern United States is an enhancement of the flow of highly polluted air northeastward from industrialized areas. This results in part from the accumulation of pollutants within the air mass, coupled with weak anticyclone surface flow, and in part from steady southwesterly winds typically found at heights of a few thousand feet to the west of the surface anticyclone as a result of the associated upper air pressure pattern. Low frequencies occur during the winter, and a secondary minimum occurs in July. Figure 3-7, sketched from data in a figure presented by Korshover, shows the geographic distribution of the average annual number of stagnation days for the 1936 to 1965 period.

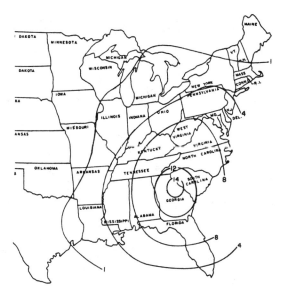

Figure 3-7. Average annual number of stagnation days, 1936 to 1965 (see Korshover[3]).

In a study of air pollution potential for the western United States, Holzworth[4] carried out a somewhat similar investigation of the occurrence of stagnating anticyclones, although the criteria and study techniques had to be different because of the irregularity of the terrain. In this investigation an occurrence was counted if it persisted for 2 days or longer. Holzworth found that these conditions were extremely rare except during winter and late fall. The average annual number of occurrences over the total study area during an 8-year period was 20, and the area having the highest frequency (about one event per year) was centered over Idaho and Wyoming.

ATMOSPHERIC CHEMISTRY AND PHYSICS

Chemical Transformation during Transport

Analyses of precipitation samples collected in North America indicate that their acidity is controlled to a great extent by the concentration of sulfate and nitrate ions. The discussion that follows, therefore, focuses on the formation of these two pollutants. Although small amounts of sulfates and nitrates are emitted directly into the atmosphere, the major sources of sulfates and nitrates in precipitation are believed to be the oxidation end products of sulfur oxides and nitrogen oxides. Near oil-burning power plants and smelters, however, direct emissions of sulfates may be of significance. The rate at which oxidation occurs between the point of emission and the receptor is therefore critical in determining the acidity of both wet and dry deposition products. Current understanding of the conversion rates of sulfur and nitrogen oxides to sulfates and nitrates and the factors affecting these rates are summarized below.

Conversion of Sulfur Dioxide to Sulfate--

Conversion of sulfur dioxide to sulfate in the atmosphere may occur as a result of two types of reactions. The first occurs as a result of photo-oxidation when all involved substances are in the gas phase (homogeneous oxidation). Direct photo-oxidation of sulfur dioxide in pure air is believed to be neglibible. If it occurs at all, the rate is probably slower than 0.03 percent per hour.[5] In polluted atmospheres, homogeneous oxidation of sulfur dioxide proceeds at a more rapid rate after gas-phase collision with strong oxidizing radicals such as HO, HO_2, and CH_3O_2. The source of these radicals is hydrocarbon-NO_x emission, which through daytime photo-oxidation produces oxidizing radicals as intermediate products.[6] The rate of oxidation is believed to depend on the initial ratio of hydrocarbons to NO_x, temperature, dewpoint, solar radiation, and the absolute concentrations of the reactive pollutants. Direct observations of reactive transients under a variety of atmospheric conditions would be needed to confirm details of the oxidation mechanism; however, such data are not available. Table 3-1 prepared from estimates presented by Barnes,[7] summarizes probable homogeneous oxidation rates for a variety of conditions. Other information compiled by the National Academy of Sciences[6] suggests rates ranging from 0.5 to 5 percent per hour for sunny summer days. Wintertime rates are expected to be lower by a factor of 2 to 5 or more because of reduced sunlight intensity.

TABLE 3-1. ESTIMATED HOMOGENEOUS OXIDATION RATES FOR SULFUR DIOXIDE

Degree of pollution	Oxidation rate (%/hour)
Pure air	Negligible (<0.03)
Natural background	<0.1
Urban mixture (containing both NO_x and olefinic hydrocarbons)	1-10

The second type of reaction involves both gaseous and liquid or solid phases and is therefore called heterogeneous. Three heterogeneous mechanisms are believed to be important in the atmospheric conversion of sulfur dioxide.

They are: (1) catalytic oxidation in water droplets by transition metals; (2) oxidation in the liquid phase by strong oxidants such as ozone and hydrogen peroxide originating from the gas-phase photo-oxidation of hydrocarbon-NO_x mixtures, and (3) surface-catalyzed oxidation of sulfur dioxide on collision with solid particles, particularly elemental carbon (soot). Unfortunately oxidation rates for these heterogeneous reactions in the atmosphere are not known. The relative roles played by the two types of reactions in removing sulfur dioxide from the atmosphere have been very approximately estimated by Friend[8] to be 10 percent by homogeneous oxidation and 90 percent by heterogeneous oxidation.

The rate reactions for homogeneous oxidation presented above are based on laboratory simulations and chemical kinetic model calculations. Two techniques have been principally relied on for verification of oxidation rates in the atmosphere. The most direct approach has been through the study of individual power plant plumes, some of which have been traced over distances up to 400 kilometers and for periods of up to 12 hours under favorable conditions. Plumes in the St. Louis, Missouri, area have shown a diurnal cycle with daytime rates varying between 1 and 4 percent per hour, and nighttime rates consistently below 0.5 percent per hour.[9] Figure 3-8, taken from Huser,[9] presents conversion rates determined from nine sampling flights at one power plant as a function of time of day.

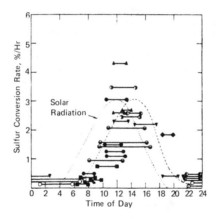

Figure 3-8. Sulfur conversion rates for the Labadie power plant plume for nine aircraft sampling flights. The points to the left of each bar are the release times and to the right, the sampling times.

Conversion rates from 0 to 5.7 percent per hour, also determined from power plant plume measurements, have been reported by Hegg and Hobbs.[9a] In these experiments, the rate was found to vary with travel time from the stack and the ultraviolet light intensity. The oxidation rate of SO_2 within a plume from a smelter located in an unpolluted area of Australia has also been determined experimentally by Roberts and Williams[9b] who concluded that oxidation proceeded only during the day. They also found the rate of oxidation to be 0.25 percent per hour when averaged over a 24-hour period. These observations were made during the winter season.

The second approach uses a regional model in which rate constants for the conversion and removal of sulfur dioxide are included as unknowns and are determined from a best-fit comparison between calculated and observed values. From a practical point of view, this regional approach has the advantage of providing rate constants that are averages over all sources included in the emission inventory and over the spatial-temporal scales of interest. Application of this technique in Europe leads to a year-round average conversion rate of SO_2 to SO_4 of 1 to 2 percent per hour.[10]

Conversion of Nitrogen Oxides to Nitrates--

The chemistry of nitrogen oxides in the atmosphere is complex and the details are not well understood. Of the various processes involved, those by which nitric oxide (NO) and nitrogen dioxide (NO_2) are converted to acidic end products are of principal concern in the development of acid rain. Other oxides of nitrogen occur only in very low concentrations. The conversion of NO_x to nitric acid takes place through a series of complicated reactions during which participating nitrogen oxides switch back and forth between various stages of oxidation and eventually end up as nitrates.

In very general terms, nitric oxide emissions are converted partially to NO_2 as a result of gas-phase reactions. Although NO will oxidize to NO_2 according to the equation

$$2NO + O_2 \rightarrow 2NO_2,$$

the rate of oxidation is highly dependent on concentration. During initial dilution with air, the conversion rate may be as rapid as 8 percent per minute.[11] For typical ambient levels of nitric oxide, however, this reaction proceeds very slowly. For example, at 0.1 ppm, its half-life is approximately 1000 hours.[12] In contrast, conversion may take place rapidly (within a matter of seconds) in a polluted atmosphere exposed to solar radiation, as it does in the formation of smog. The key to this rapid conversion lies in sequences of reactions between transient species and other reactive molecules such as, carbon monoxide, hydrocarbons, and aldehydes. During daytime conversion, an important photochemical cycle takes place in which both the production and destruction of ozone occurs. As part of this process NO_2 is converted to nitric acid vapor and NO, and NO_2 may be absorbed into existing particles.

Most of our understanding of the oxidation processes and photochemical reactions taking place in the atmosphere has come through laboratory studies involving chemical and kinetic modeling. Key information needed to tie these studies firmly to reactions within the atmosphere is still missing.[11] Furthermore, it will be very difficult to obtain this information directly because many of the reacting substances are transient, secondary pollutants that occur only in very small concentrations. There are numerous possible paths by which nitrate salts may be formed from gaseous nitric acid. Some of these involve homogeneous processes such as the direct capture of gaseous nitric acid by gaseous ammonia, which leads to the formation of ammonium nitrate. Others, also believed to be of importance, include heterogeneous reactions occurring during the absorption and accumulation of acids and other substances by aerosol droplets and cloud water.

Because of the complexity of the chemical processes involved in the production of acidic products from nitrogen oxides in the atmosphere and the spatial and temporal variations of key parameters controlling these processes, rates of conversion of nitrogen oxides to nitrates can be expected to vary greatly. Thus, Haagen-Smit and Wayne[12] state that these processes may take hours or days. Barnes[7] reports that precise conversion rates are not known but are thought to vary with season. Barnes further states that the atmospheric lifetime of nitrogen compounds is greater than that for sulfur oxides. The degree of uncertainty associated with the conversion of NO to NO_2 and NO_2 to nitric acid is obviously very great.

Removal of Pollutants by Precipitation

For discussion purposes, the processes by which pollutants are removed from the atmosphere during cloud growth and precipitation are frequently separated into two groups. Processes that occur within the cloud are referred to as rainout and begin with the condensation of water vapor on cloud condensation nuclei (CCN). Many of these nuclei are believed to be sulfate particles that have been formed as a result of the gas-to-aerosol conversion of sulfur dioxide emissions. Condensation is followed by droplet growth, during which various pollutants dissolve in the droplets, undergo chemical changes, and begin their descent to the ground in falling precipitation. The removal of pollutants below the cloud base by this precipitation is called washout. Of the two processes, rainout is the more complex, and no adequate quantitative theory covering all aspects of it exists.[13] However, an appreciation of its complexity can be gained by considering the process of cloud formation.

Clouds are dynamic systems within which water vapor is condensing, evaporating, and recondensing. During these cycles large volumes of air are processed by the clouds, but it has been estimated that even in large storms, only about half of the water vapor passing upward through the cloud base is returned to the ground as precipitation. The initial condensational growth on CCN can produce droplets about 10 μm in radius in 5 to 10 minutes; thereafter growth by condensation is extremely slow.[14] Growth of droplets to precipitable size occurs principally by the collision and coalescence of smaller cloud droplets. Within the clouds, pollutants are absorbed by droplets, participate in chemical reactions, and frequently return to cloud-free air at higher elevations following droplet evaporation. Thus the acidity of droplets depends not only on the ionic species present within the droplet but also on their dilution within the droplet. This is a function of where the droplet is in its condensation-evaporation cycle. Pollution may be cycled through the aqueous phase many times before being deposited on the ground by precipitation.[15] Cloud systems vary in horizontal and vertical dimensions and in internal vertical motions. Also, the life history of individual droplets varies greatly from cloud type to cloud type as well as from drop to drop within the same cloud.

Attempts have been made to calculate the relative importance of rainout and washout to the concentration of ionic species in precipitation. The general approach followed has been to calculate the contribution produced by washout and to ascribe the remaining species found in the precipitation to in-cloud scavenging processes. As pointed out by Marsh,[13] the washout of SO_2 by rain falling through a uniform concentration of SO_2 is a function of: (1) the size spectrum of droplets and hence the rate of precipitation, (2) the initial pH of the rain, (3) the height of the SO_2 concentration, and (4) the absolute magnitude of the SO_2 concentration. Furthermore, the joint washout of SO_2 with other pollutants differs from that of SO_2 alone. In particular, the presence of NH_3 hastens the absorption of SO_2 and its conversion to sulfate.[16] Calculations by Marsh[13] suggest that the scavenging of SO_2 gas and particulates below cloud base and in-cloud scavenging contribute about equally to the concentration of sulfates in precipitation. Also of interest are network data from MAP3S that indicate that sometimes as much as 30 percent of the total sulfur in rainwater may be a result of direct scavenging of SO_2. Calculations of this type are useful in developing an understanding of the physical and chemical mechanisms involved. When the mechanisms are understood, major simplifications will be needed before the entire process can be satisfactorily simulated in regional models.

Similar mechanisms are involved in the scavenging of NO_x, nitrates, and other pollutants from the atmosphere but the details of the individual processes are even less well understood.

Effects of Stack Height

For a period of years, one of the most popular methods for reducing the ground-level concentration of pollutants in the vicinity of power plants and industrial complexes has been the use of tall stacks. For example, the average height of stacks built for coal-burning power plants in 1969 was 609 feet compared to 243 feet for those built in 1960, and numerous stacks as high as 1000 feet exist today. Although the tall stack by itself does not reduce the amount of pollutants introduced into the atmosphere, the additional dilution experienced by the pollutants under most meteorological conditions is frequently sufficient to keep ground-level concentrations below ambient standards. The overall effect of using a tall stack on atmospheric chemistry and the ground-level concentrations of transformed pollutants is not yet completely clear. Nonetheless, it is probable that the contribution to regional-scale pollution problems is somewhat enhanced by the use of tall stacks because release above the mixing height occurs more frequently with increasing stack height. When released above the mixing height, emissions may not reach the earth's surface for many miles. During this period of transport they may experience chemical transformations but are not subject to the substantial losses by dry deposition that low-level emissions experience close to the source.

An attempt to estimate the effect of stack height on the far field wet and dry deposition of sulfur over Europe by modeling has been reported by Fisher.[18] Two calculations were performed: The first assumed that half the sources in Europe are at a height of approximately 300 meters (984 feet) and that the remaining sources were at approximately 30 meters (98 feet); the second calculation assumed that all of the sources were low-level sources. The effect of changes in source height on wet deposition was found to be less than about 10 percent. The effect on dry deposition was not reported. Fisher's model takes a statistical approach to obtain the long term deposition pattern. The model is being developed to provide a fairly rapid and efficient method of determining the pattern of total and wet annual deposition of sulfur over Europe. The results are considered no more accurate than a factor of two, but have been found to be in good agreement with the results of models based on trajectory computations.

The 1977 amendments to the Clean Air Act recognize the limitations of the tall stack approach to the control of pollution. Section 123 states that "the degree of emission limitation required for control of any air pollutant under an applicable implementation plan under this title shall not be affected in any manner by (1) so much of the stack height of any source as exceeds good engineering practice." Good engineering practice is interpreted to mean "the height necessary to insure that emissions from the stack do not result in excessive concentrations of any air pollutant in the immediate vicinity of the source as a result of atmospheric downwash, eddies and wakes which may be created by the source itself, nearby structures or nearby terrain obstacles"-- and is usually taken to be two and a half times the height of the source.

Relative Contributions of Wet and Dry Deposition

Until recently, wet deposition was considered to be the chief mechanism by which acidic substances are removed from the atmosphere and deposited on the surface of the earth. Within the last few years, however, consideration of more extensive observations, regional modeling results, and mass balance calculations has led to the view that dry deposition is probably of equal importance over western Europe and North America.[4,19] Because of early interest in sulfur dioxide and sulfates and the consequent existence of substantial data bases, most of the estimates carried out to date have been for sulfur compounds. Although preliminary, these estimates are useful in bringing to attention the role played by dry deposition. Many of these results have been expressed in terms of total sulfur rather than as the separate sulfur dioxide--sulfate constituents that are of greater interest in evaluating contributions to acid deposition. To illustrate the results obtained by current techniques, results determined for the eastern United States and Canada by two different approaches are presented below.

Shannon, in a recent report,[20] has, using the ASTRAP model, presented example calculations that permit the relative importance of the wet and dry deposition of sulfur to be compared. The calculations covered three periods: July-August 1974, January-February 1975, and July-August 1975. The model was run twice--once with emissions from the eastern United States only and once with emissions from eastern Canada only. The relative contributions of the two deposition processes are shown in Table 3-2. In both model runs, wet

TABLE 3-2. RELATIVE CONTRIBUTIONS OF WET AND DRY DEPOSITION
(Based on ASTRAP Model--Shannon[20])

Period	United States		Canada	
	Dry	Wet	Dry	Wet
U.S. Emissions Only				
Jul-Aug, 1975	1	1.1	1	2.6
Jul-Aug, 1974	1	1.2	1	3.3
Jan-Feb, 1975	1	1.7	1	3.1
Average	1	1.3	1	3.0
Canada Emissions Only				
Jul-Aug, 1975	1	1.1	1	1.7
Jul-Aug, 1974	1	1.2	1	1.9
Jan-Feb, 1975	1	1.5	1	1.4
Average	1	1.3	1	1.7

deposition was slightly more effective than dry deposition in removing sulfur from the atmosphere over the United States (1.3 to 1). Over Canada, however, wet deposition proved to play a considerably greater part, particularly for emissions entering from the United States (3 to 1).

The second example is provided by Galloway and Whelpdale,[21] who have developed an atmospheric sulfur budget for eastern North America (see Table 2-3). Their estimates of wet deposition were made from observations throughout the eastern United States and Canada. Dry deposition was calculated from representative air concentrations of SO_2 and $SO_4^=$ and deposition velocities. The authors estimate the accuracy of their dry deposition estimates to be approximately ±90 percent for both Canada and the United States. Using their results, the ratios of wet to dry deposition over the two regions are: for the eastern United States, 0.76 to 1; and for eastern Canada, 2.5 to 1. Consideration of possible errors introduced into the wet deposition estimates by the use of monthly collections led Galloway and Whelpdale to suspect that their estimates of wet deposition might be high by up to 30 percent. According to their estimates, the dry deposition of sulfur as SO_2 exceeds that as $SO_4^=$ by a factor of about four in Canada and by a factor of about five in the United States.

A table of representative annual average wet and dry deposition rates for various parts of the world was prepared during the workshop held at the time of the International Symposium on Sulfur in the Atmosphere held in Dubrounik, Yugoslavia, in September 1977.[19] This table is reproduced here as Table 3-3. Ranges represent spatial variations for each area.

TABLE 3-3. REPRESENTATIVE ANNUAL AVERAGE WET AND DRY DEPOSITION RATES

Location		Excess sulfate in precipitation (mg S ℓ^{-1})	Wet deposition rate (g S $m^{-2}y^{-1}$)	Dry deposition rate (g S $m^{-2}y^{-1}$)
Heavily industrialized areas	North America	3-?	0.1^a-3	?
	Europe	3-20	2-4	3-15
Rural	North America	0.5-2	0.1-2	0.2-2.6
	Europe	0.5-3	0.2-2	0.5-5.0
Remote	North Atlantic	0.2-0.6	0.1-0.3	0.04-0.4
	Other Oceans	0.04	0.01^a-0.2	<0.1
	Continents	0.1	0.01^a-0.5	0.4

^{a}Low deposition rates result from low precipitation.

Calculations of deposition rate are usually performed either by multiplying the air concentration at a specified height by an experimentally derived parameter called the <u>deposition velocity</u>, or in certain situations by a mass balance approach. Deposition velocities have been derived for a few typical locations, but it is important to realize that the deposition process depends on many factors. Dry deposition rates are particularly susceptible to atmospheric parameters such as stability, turbulence, and windspeed, and to surface parameters such as physical roughness, vegetation, and chemical characteristics. An appreciation for the wide range of deposition velocities possible over various types of vegetation and soil can be gained from the estimates for SO_2 presented in Table 3-4, a and b. These tables were also developed during the workshop at the International Symposium on Sulfur in the Atmosphere.[19] Dry deposition velocities for sulfate particles are usually assumed to be about 0.1 cm s^{-1}, but some observations have indicated that values five times as great might be more appropriate in light and moderate winds over natural grassland.[22]

Variability of Acidity within and among Storms

Detailed observations of precipitation pH show that there are frequently very large variations in acidity not only from storm to storm at individual sites but also spatially and temporally within the same precipitation event. Differences found within an individual storm might conceivably result from spatial and temporal differences in the cloud structure (e.g., vertical motions, droplet sizes, vapor pressures) and the associated variations in rainout efficiency, plus differences in the efficiency with which acidic substances are washed out from below the cloud. On the other hand, these

TABLE 3-4a. DEPOSITION VELOCITIES OVER VEGETATION

Vegetation		Height (m)	Range of V_g (cm s^{-1})	Typical V_g[a] (cm s^{-1})
Height	Example			
Short	Grass	0.1	0.1 to 0.8	0.5
Medium	Crops	1.0	0.2 to 1.5	0.7
Tall	Forest	10.0	0.2 to 2.0	Uncertain

[a] These values were obtained in a humid climate. Much smaller values are likely in arid climates

TABLE 3-4b. DEPOSITION VELOCITIES OVER SOIL

Soil type	pH	State	Range of V_g (cm s^{-1})	Typical (cm s^{-1})
Calcareous	≥ 7	Dry	0.3 to 1.0	0.8
Calcereous	≥ 7	Wet	0.3 to 1.0	0.8
Acid	~4	Dry	0.1 to 0.5	0.4
Acid	~4	Wet	0.1 to 0.8	0.6

Note: As yet no information is available to assess V_g on desert sand or lateritic soils.

differences in pH might also be caused by localized source differences of either acidic substances and precursors or of neutralizing material. This section reports some of the variations that have been observed to indicate the probable complexity of the mechanisms involved.

Deposition as a Function of Precipitation Amount--
Pack[15] observes that most of the formulations for wet deposition assume, and much of the data indicate, that removal rates are proportional to precipitation rates. He also states that the concentration of various ions (e.g., H^+ and $SO_4^=$) is often highest in light precipitation, as shown in Figure 3-9. Pack concludes with the statement that "speaking loosely, it appears that the higher ion concentrations in light rain tend to balance the lower values in heavier precipitation so that the total material deposited on the ground is much less variable than would otherwise be the case."

Spatial Variability within a Metropolitan Network--
In Figure 3-10, Semonin[24] shows a precipitation weighted mean pH pattern based on measurements at 81 sites within an area of 1800 square kilometers in the St. Louis area. The data were collected during 22 precipitation events as part of the METROMEX (Metropolitan Meteorological Experiment). The areal mean of the data is pH 4.9 with a range of values from 4.3 to 6.8. Semonin suggests that these results may reflect a rapid response of precipitation formation

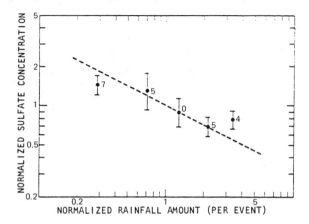

Figure 3-9. Variation of $SO_4^=$ concentration with precipitation amount, MAP3S station data.[53]

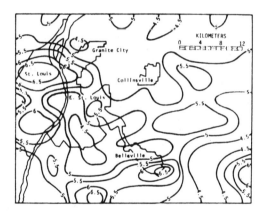

Figure 3-10. The rainfall weighted mean pH pattern.

processes to scavenging of either water-soluble alkaline or acid aerosol from unevenly distributed local sources. Semonin also presents network measurements for two of the individual storm events, one with a weighted mean of pH 4.8 and a range of values from 3.4 to 8.2, and the other with a mean of pH 6.4 and a range from 5.2 to 7.9. Even knowing the locations and general character of industrial effluents it was impossible for him to determine uniquely the causes of acid or alkaline rainfall, however.

Temporal Variation within a Storm--

The temporal variation of pH in three rain storms at Gainesville during 1976 as reported by Hendry[25] is shown in Figure 3-11. All plotted values except the final one for each storm represents a 5-minute average. In each storm, the least acidic value was observed at the start of the storm. Figure 3-12 shows a similar plot of pH measurements made during two rainfall events in Tuscon, Arizona, in the summer of 1977. The observations made in Series D covered a 13-minute period during which 4.5 mm of rain fell; Series H covered a 33-minute period during which 10.7 mm of precipitation fell. In both of these examples, the least acidic values also occurred at the beginning of the storm.

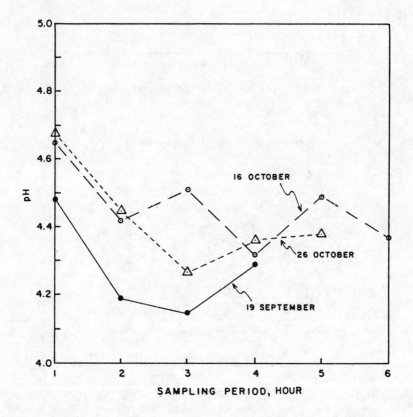

Figure 3-11. Variation of pH during three storms in Gainesville, Florida.

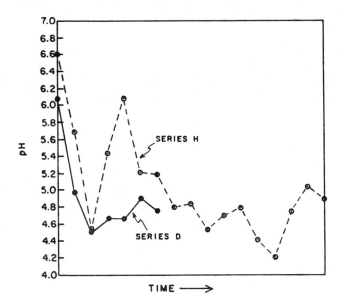

Figure 3-12. Variation of pH during two storms in Tucson, Arizona.

In contrast, Raynor[27] has reported measurements from two storms at the Brookhaven National Laboratory, one of rain and one of snow, in which the pH rose quite steadily during the period of precipitation. In other storms, he reported that the cleansing effect of precipitation was not evident, either because the rates of rainfall were too low or the air was too clean at the beginning of precipitation.

Topographic Impacts

Topographic features such as the existence of bodies of water and the presence of mountain ranges influence the amount, distribution, and acidity of precipitation. Because many of the effects of acid rain are cumulative, the total annual precipitation is important as well as its average acidity. Orographic precipitation is produced when moisture-laden air rises and cools while passing over a topographic barrier. It is an important factor in intensifying rainfall on windward slopes. Precipitation on the leeward side of a range tends to be diminished or absent because of downslope motion and the resulting warming of the air mass. Additionally, orographic barriers hinder the passage of storms and frontal systems and promote convection by differential heating along the slopes. The frequent occurence of snow immediately south of the Great Lakes when cold northerly winds flow over open water in the early part of winter is a well-known illustration of the influence of a moisture source on local precipitation patterns.

The rugged north-south mountain ranges along the west coast of North America prevent the moderately warm and humid Pacific air masses from entering the central portion of the continent without major modification. The prevailing westerly winds and the impact of storms from the Pacific result in heavy orographic precipitation along the coast, particularly north of 40° latitude, and lesser amounts east of the mountains. The Rocky Mountains play a similar but less marked role in determining precipitation patterns because of the lower average moisture content of the air masses. They also provide a barrier to the occasional storm systems that move against the mountains from the south and east. Although the mountains near the East Coast are not a very effective climatic barrier, they do have an important influence on the weather under certain meteorological regimes.

A mountain barrier, such as the Adirondacks, which is centrally located with respect to storm tracks can be expected to enhance precipitation within the mountains. In the case of the Adirondacks, much of this precipitation is very likely to be quite acidic because there are sources of suspected precursor pollutants upwind of the mountains for nearly all wind directions. The long axes of the pH isopleths for the region, such as those in Figure 3-25 presected later, do fall approximately along the mountain range and also show the lowest pH values in the vicinity of the mountains, as would be expected. However, as pointed out later in this section, the monitoring data available for analyses in the past for the eastern United States have been quite sketchy. A rigorous investigation of the occurrence of wet (or dry) deposition maxima and their association with topographic features will require data from a comprehensive monitoring network coupled with an analysis of air parcel trajectories.

An illustration of the extent to which a mountain range can remove acidic pollutants from an air mass has been given by Likens et al.[28] In this example, the pH of precipitation from air masses arriving from the British Isles increased from an average of 4.2 at the Norwegian coast to 4.6 or 4.7 after passing over mountainous areas with an elevation of 1000 meters (3280 feet) or so. Presumably, precipitation on the leeward side of the mountains was being produced within an airmass already partially cleansed by earlier precipitation on the windward side.

REVIEW OF REGIONAL TRANSPORT AND DEPOSITION MODELING

In this section, three regional modeling studies carried out in the northeastern U.S. and eastern Canada are reviewed. The status of regional modeling is summarized, and the areas of uncertainty are assessed.

The PNL Regional Model

Battelle, Pacific Northwest Laboratories (PNL), has developed a regional transport model for sulfur oxides that determines monthly and annual concentrations of sulfur dioxide (SO_2), sulfate ($SO_4^=$) and the total monthly and annual wet deposition of sulfur. The PNL model as described by Powell et al.[29] simulates transport of large point source plumes by following their trajectories with reference to a fixed grid. The wind field used is obtained by averaging National Weather Service (NWS) radiosonde data in a 100-to-1000-m layer. The average wind vectors are interpolated to generate a representative

wind field. The fields are then interpolated every hour between 12-hour
observation intervals. Trajectories originate hourly at each point source
and calculations of diffusion, transformation, and wet and dry deposition are
made using mass conservation equations. The mass of pollutant is added to
the appropriate grid cell to determine time-averaged concentration fields.

The transformation of SO_2 to $SO_4^=$ is assumed to be linear. The transformation rate is constant spatially, but can vary with time. Dry deposition of SO_2 and $SO_4^=$ are accounted for by deposition velocities, which are assumed to be constant. Wet removal of SO_2 is directly proportional to the hourly precipitation rate, whereas wet removal of $SO_4^=$ is based on a formulation by Scott[30] and is proportional to the hourly precipitation rate raised to the 0.625 power. Mixing height is given a sinusoidally varying cycle to represent the buildup of the daytime mixed layer and the nocturnal stable layer.

The PNL regional model was applied to the eastern United States and eastern Canada, Figure 3-13. The critical model parameter values used in the study are listed in Table 3-5. In the most recent version of the model, the SO_2 transformation rate is increased to 10 percent per hour during precipitation events to account for in-cloud processes, as suggested by Scott and Laulainen.[32] The SO_2 and $SO_4^=$ wet removal rate constants are based on work by Dana et al.[33]

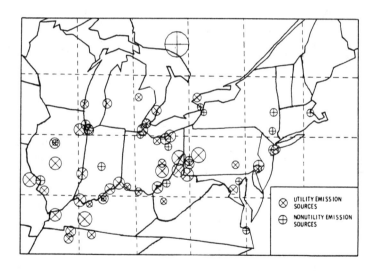

Figure 3-13. Area of PNL model calculations
and emission source locations
(McNaughton).[31]

TABLE 3-5. REGIONAL MODEL PARAMETER VALUES FOR EASTERN UNITED STATES AND CANADA TRANSPORT SIMULATIONS

	PNL	EURMAP-1	ASTRAP
SO_2 transformation rate (%/hr)	Day 2.0 Night 0.25 Rain event 10.0	1.0	Diurnal cycle Summer 0.25-3.0 Winter 0.1-1.5
SO_2 dry deposition velocity (cm/s)	1.0	1.0	Summer 0.1-0.8 Winter 0.1-0.7
$SO_4^=$ dry deposition velocity (cm/s)	0.1	0.2	Summer 0.1-0.8 Winter 0.1-0.6
SO_2 wet removal rate (%/hr)	0.5 P(t)	28 P(t)[a]	$100(h/4)$ $1/2$ $h^b \leq \frac{4\ mm}{6\ hr}$
$SO_4^=$ wet removal rate (%/hr)	23.2 P(t)	7 P(t)[a]	100 $h^b > \frac{4\ mm}{6\ hr}$
Mixing depth (m)	Day 600 - 1200 Night 200 - 400	Winter 1150 Spring 1300 Summer 1450	2100 10 levels

[a]Precipitation rate in mm/hr.
[b]Precipitation rate in mm/6 hr.

Transport during the 2 months August and October 1977 were simulated using 1973-1974 large point source sulfur emission data from NEDS[34] and the Federal Power Commission.[35] According to the modelers, they account for approximately 65 percent of the total emissions in the region; remaining emissions are from area sources and small point sources.

Model statistics for two cases are compared to observations in Table 3-6. Case 1 takes into account increased transformation during precipitation events. In both cases the PNL regional model underpredicts SO_2 and $SO_4^=$ monthly average concentrations. The modelers attribute this underprediction to the fact that only 65 percent of the total emissions are used.

The sulfur wet deposition statistics are compared to data taken at four stations in the U.S. Department of Energy's MAP3S precipitation chemistry network (Dana[37]) in Table 3-7. Without the increase in transformation during rain events, Case 2, the PNL model underpredicts sulfur deposition by as much as a factor of 3. Case 1 results are in much better agreement, perhaps agreeing too closely with data considering that only 65 percent of the total emissions are being transported. The observed and Case 2 predicted $SO_4^=$ concentration fields for August 1977 are shown in Figure 3-14.

TABLE 3-7. PNL MODEL EVALUATION STATISTICS FOR MONTHLY WET DEPOSITION OF SULFUR (AS $SO_4^=$ IN g m^{-2} h^{-1} (McNAUGHTON AND SCOTT)[36]

Station	August 1977			October 1977		
	Observed	Case 1	Case 2	Observed	Case 1	Case 2
Whiteface, NY	5.96	6.90	4.83	3.45	2.03	1.30
State College, PA	5.35	6.71	3.67	4.28	4.34	1.47
Charlottesville, VA	5.02	6.92	3.41	0.69	0.66	0.31
Ithaca, NY	1.56	0.59	0.49			
Mean $\frac{\text{Predicted}}{\text{Observed}}$		1.04	0.62		0.85	0.39

The PNL model is still in a testing and developmental stage and no conclusions can be reached from the preliminary results. The model is promising and may serve as a useful tool to judge the impacts of proposed industrial development scenarios.

The SRI EURMAP-1 Regional Model

SRI International developed a trajectory-type regional air pollution model, EURMAP-1 (European Regional Model Air Pollution), for the Federal Environmental Agency of the Federal Republic of Germany.[38] EURMAP-1 calculates long-term average concentrations, and dry and wet deposition of SO_2 and $SO_4^=$. This model was adapted for the region of eastern North America as shown in Figure 3-15.

TABLE 3-6. SUMMARY OF PNL MODEL EVALUATION STATISTICS FOR CONCENTRATIONS (McNAUGHTON AND SCOTT)[36]

Summary of model evaluation statistics for air concentrations	Ambient $SO_4^=$ concentration				Ambient SO_2 concentration			
	August 1977		October 1977		August 1977		October 1977	
	Case 1	Case 2	Case 1	Case 2	Case 1	Case 2	Case 1	Case 2
Mean observed concentration ($\mu g\ m^{-3}$)	10.25	10.25	6.24	6.24	14.97	14.97	21.70	21.70
Mean predicted concentration ($\mu g\ m^{-3}$)	8.09	5.73	5.74	4.46	11.56	13.29	11.85	12.92
Correlation of observed to predicted concentration	0.65	0.65	0.39	0.59	0.57	0.56	0.75	0.76
Variance observed data	3.41	3.41	1.82	1.82	54.75	54.75	159.41[a]	155.41
Variance predicted data	20.52	20.29	2.58	1.66	38.92	44.66	39.54	39.96
Maximum observed concentration ($\mu g/m^3$)	14.60	14.60	8.92	8.92	30.13	30.13	69.17	69.17
Maximum predicted concentration at monitoring sites ($\mu g/m^3$)	21.83	14.90	8.74	7.31	23.35	25.20	23.99	45.69
Frequency of concentrations within a factor of (%): 2	70	59	100.00	89	73	78	62	62
3	88	72	100.00	99	99	86	79	89

[a] Apparent bias by a local source influence at Johnstown, PA.

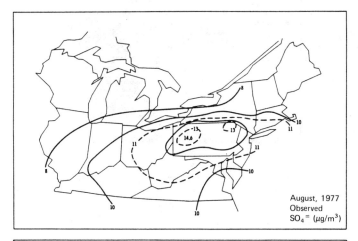
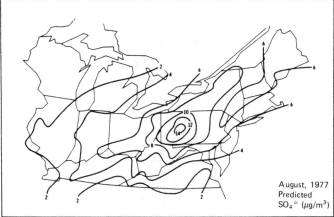

Figure 3-14. Comparison of observed $SO_4^=$ concentration fields to PNL model predictions (Case 2) for August 1977.

Figure 3-15. EURMAP-1 modeling domain and EPA regions. (Bhumralkar et al.)[39]

The basic principles of the model are illustrated in Figure 3-16. Puffs of $SO_2/SO_4^=$ are emitted at equal time intervals from all sources. The puffs are assumed to be well mixed in the horizontal and vertical. The transport of the puffs is then calculated using the mixed layer wind field.

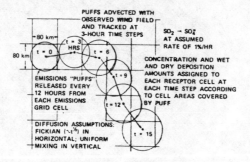

Figure 3-16. Emissions puff advection and diffusion scheme used in EURMAP-1. (Bhumralkar et al.)[39]

During transport, pollutant mass is removed from a puff by dry deposition at a user-specified rate. Wet removal of SO_2 and $SO_4^=$ during rain events occurs at a constant rate proportional to the rainfall rate. Transformation of SO_2 to $SO_4^=$ is treated linearly. Sulfur dioxide and sulfate concentrations for each puff are determined over time using mass balance equations. The mass is added to the appropriate grid cell to determine time average grid concentration fields.

The wind field is determined by objective analysis of available upper-air observations at the 850-mb level (approximately 1500 m above mean sea level). The resulting field wind speeds are decreased by 1/4, and the wind directions are rotated 15° counterclockwise to account for surface layer friction effects. The wind fields are then interpolated every 3 hours between 12-hour data intervals.

The parameter values listed in Table 3-5 were used in the model simulations. With the exception of wet removal, the values are similar to those used in the PNL model. The wet removal rates are quite different, indicative of the uncertainties involved in the present state of wet removal modeling. The EURMAP-1 model does not include any special treatment of incloud transformation.

In the model simulation, $SO_2/SO_4^=$ emissions consisted of the 1977 emission data base prepared for the SURE program by GCA Corporation. The months of January and August 1977 were chosen for the analysis, and the results were compared with SURE, NEDS, and SAROAD air quality data. The calculated and measured concentration fields of $SO_4^=$ are compared in Figures 3-17 and 3-18. In January, EURMAP-1 predicts high sulfate in the northeastern states and relatively low values elsewhere. The observed concentration field is similar in the East but measured values are much higher than predicted in the Midwest. The model results for August are in better agreement with observations. Wet deposition patterns were calculated (Figure 3-19), but, unfortunately, they were not compared with data.

The EURMAP-1 approach is similar to the PNL model formulation; the biggest difference is in how they parameterize wet deposition. The month-averaged $SO_4^=$ concentration field calculated by EURMAP-1 for August 1977, Figure 3-18, shows a distribution similar to results of the PNL, Figure 3-14. The magnitude of the PNL concentrations are closer to the observed values.

The ASTRAP Regional Model

Argonne National Laboratory has developed the Advanced Statistical Trajectory Regional Air Pollution (ASTRAP) model for sulfur under the MAP3S program initiated by the U.S. DOE. The model was developed by Sheih[40] and was based on a concept introduced by Drust.[41] As its name suggests, this model takes a statistical approach to long-term regional modeling rather than to the event simulation technique adopted by PNL and EURMAP-1. The ASTRAP model is based on the assumption that for long-term averaging (periods of one month or longer) horizontal and vertical dispersion processes can be separated. Based on this assumption, the model consists of three independent subprograms. First, the long-term horizontal dispersion statistics (mean

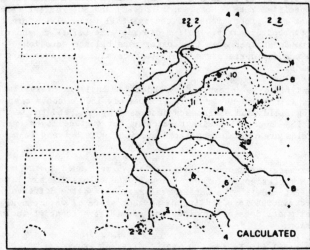

Local maximum values shown apply at points marked by plus signs

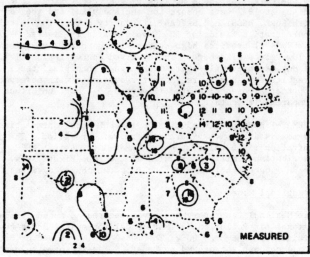

$SO_4^=$ concentrations ($\mu g/m^3$) for January 1977

Figure 3-17. EURMAP-1 calculated $SO_4^=$ concentration field and the corresponding measured field for January 1977. (Bhumralkar et al.)[39]

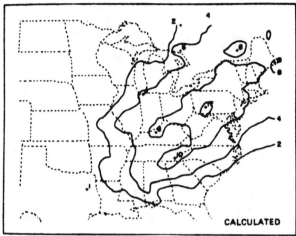
Local maximum values shown apply at points marked by plus signs

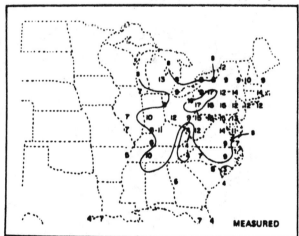
$SO_4^=$ concentrations ($\mu g/m^3$) for August

Figure 3-18. EURMAP-1 calculated $SO_4^=$ concentration field and the corresponding measured field for August 1977. (Bhumralkar et al.)[39]

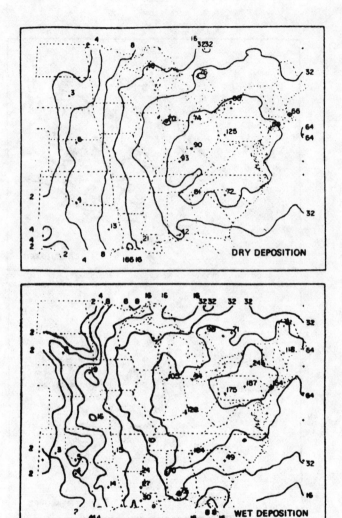

Local maximum values shown apply at points marked by plus signs

Figure 3-19. EURMAP-1 calculated dry and wet deposition of $SO_4^=$ (mg m^{-2}) for 1977. (Bhumralkar et al.)[39]

position and spread as a function of plume age) for simulated tracers released from a grid of virtual sources are calculated. Individual puffs are not tracked; instead the month-long emissions are represented by the dispersion statistics. During this step of the modeling procedure, statistics of tracer removal by precipitation are also evaluated. Next, a vertical dispersion subprogram numerically integrates the standard one-dimensional diffusion equation allowing for surface dry deposition and transformation of SO_2 to $SO_4^=$. Finally, the normalized results of the first two subprograms are used to produce fields of SO_2 and $SO_4^=$ concentrations and wet and dry deposition of total sulfur.

The ASTRAP model parameter values for the SO_2 to $SO_4^=$ transformation rate and wet and dry deposition rates are presented in Table 3-5. Seasonal and diurnal variations in transformation rate as noted by Husar et al.[42] are taken into account. Based on the work of Wesely and Hicks[43] and Wesely et al.[44] there are seasonal and diurnal cycles in the deposition velocities of SO_2 and $SO_4^=$ produced by stability and plant stomatal activity. Sulfate deposition velocities are on the same order of magnitude as SO_2 velocities rather than an order of magnitude less as in other modeling studies.

Wet removal is taken into account using the scavenging ratio approach. This method relates wet deposition to the ratio of field measurements of concentration of pollutant measured in the air to that measured in rainfall at the same time. Argonne National Laboratory has found scavenging rates to be a relatively constant value,[45] and sulfur deposition by wet processes is found to be a function of the half power of the amount of precipitation.

The mixed layer is assumed to be 2100 m in depth and is divided into 11 layers for the vertical numerical intergration. A wind field is developed at a specified level in the atmosphere based on NWS data. The 1000 m level winds are used in the winter and 1800 m level winds are used in the summer. Winds are interpolated between data points using a radius of influence inverse squared relationship.

Preliminary model runs have been made in the eastern United States and Canada using 1974 and 1975 meteorological data. The emission inventory consisted of both point and area sources emissions in the eastern United States and Canada. The model results were then compared with measurements from the SURE data network for 1977 and 1978. The average two-month summer and winter sulfate fields are compared with data in Figures 3-20 and 3-21. As these figures show, there are major discrepencies, particularly in the West. It must be kept in mind, however, that meteorology for a different year was used in the model. The ASTRAP simulations of wet deposition of total sulfur were scaled to a one-year period and compared in Figure 3-22 with observations during 1977 of annual accumulations of sulfate in precipitation, expressed as total sulfur.[21] There is some general agreement, but the data show a more complex distribution than that indicated by the ASTRAP model results. On an annual basis, an estimated 5.4 million metric tons were deposited on the eastern United States. Wet and dry removal were approximately equally important. By season, dry deposition was equal to wet deposition in the summer, but wet removal was approximately twice dry removal in the winter.

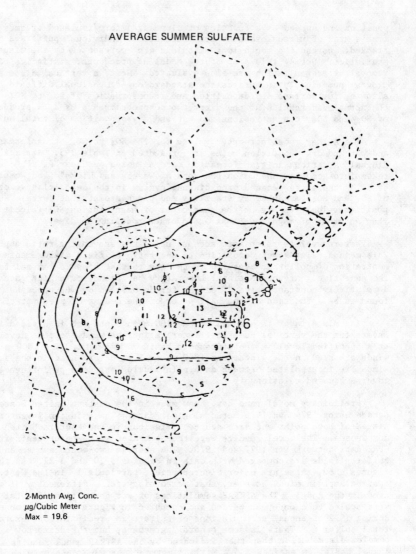

Figure 3-20. Comparison of August 1977 SURE average sulfate measurements (numbers) with ASTRAP simulations (isopleths) using July-August 1975 meteorology. (Shannon).[20]

Figure 3-21. Comparison of Jan-Feb 1977 SURE average sulfate measurements (number) with ASTRAP simulations (isopleths) using Jan-Feb 1975 meteorology. (Shannon).[20]

120 Acid Rain Information Book

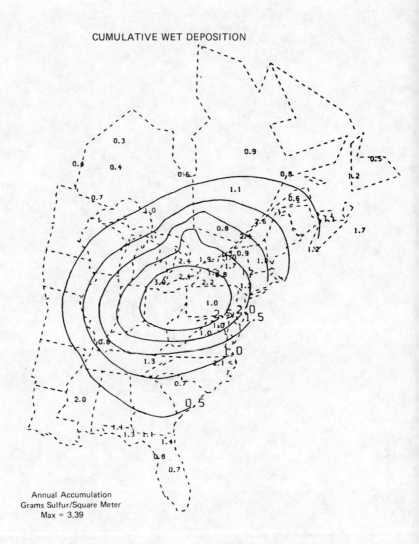

Figure 3-22. Comparison of cumulative sulfate in rain, expressed as total sulfur, for 1977 with ASTRAP simulations (isopleths) (Galloway and Whelpdale).[21]

The ASTRAP model $SO_4^=$ distribution of August 1977 is similar to the PNL and EURMAP-1 model results shown in Figures 3-14 and 3-18. The $SO_4^=$ values compare as well with data as the PNL model. At this very early stage in model development and comparison, the ASTRAP model shows as much potential as that of the PNL and EURMAP-1 approach.

Summary of State-of-the-Art Regional Modeling

Regional transport and deposition of pollutants in the northeastern United States and eastern Canada have been simulated by the PNL, EURMAP-1, and ASTRAP models. All three models use objectively generated, regional-scale wind fields based on interpolation of available upper-air data to simulate transport and diffusion of atmospheric emissions. The PNL and EURMAP-1 models are based on a trajectory approach. The ASTRAP model, instead of simulating day-to-day transport events, uses a statistical approach to determine the location of long-term mean plumes. In all three models, horizontal transport is calculated using an hourly updated windfield that is assumed to be representative of the entire mixed layer.

At this time, because of a lack of understanding of the complex chemical processes taking place, particularly within clouds, and a limited knowledge of the wet removal processes, the models rely heavily on empirical relationships to describe transformation and removal processes. The greatest disagreement among model parameterizations exists in the wet removal mechanisms. The transformation and deposition values used by the three models are compared in Table 3-5.

The models have limited preliminary simulations of SO_2 and $SO_4^=$ transport and deposition. At the present, the nitrogen cycle has not been simulated, and models are in the testing and refining stage. A limited amount of comparison of model-calculated, monthly averaged SO_2, $SO_4^=$, and total sulfur deposition to observation has been performed to evaluate the performance of the models.

The Areas of Uncertainties and Needed Improvement in Regional Transport and Deposition Modeling

As in all air pollution modeling, prescription of appropriate wind fields is crucial to obtain correct transport patterns. All studies at this time have relied on objective analysis of available NWS wind data. Unfortunately, the NWS network is designed for use in synoptic- (continental) scale modeling, and the spatial scale is not fine enough to capture the variation from synoptic-scale motion that is important in regional-scale modeling. The alternative is to use the basic thermodynamic and hydrodynamic equations to predict atmospheric behavior using real initial and boundary conditions. This work has been performed by Anthes,[47] but it is an expensive and time consuming task.

Computational and monetary restraints require that the transport and diffusion models be simple; it is hoped that the most important processes are taken into account. The atmosphere is a very complex system, however, and further evaluations may show that more sophisticated models are required to obtain meaningful results. For example, three layers might be required in the vertical: two levels in the mixed layer might be needed to account for differential transport at night of the upper part of the daytime mixed layer and the nocturnal inversion layer, and a third level above the mixed layer might be needed to track pollutants that are pumped in and out of the mixed layer by convection.

Because of a lack of data and physical understanding, the chemistry in regional models has been reduced to linear decay of SO_2 to $SO_4^=$; the nitrogen cycle has not been addressed at all. The actual chemical transformation processes include a wide variety of reduced sulfur and nitrogen species, and the chemical transformation of SO_2 and NO/NO_2 depends on temperature, relative humidity, photochemical activity, time of year, and concentration of other pollutants.

Dry deposition has been treated by the deposition velocity approach. Usually it is assumed to operate on the average concentration of the whole mixed layer, but it is a near-surface process. Consequently it is dependent on the effects of the vertical concentration profile, stability, and surface roughness, effects that have only been treated empirically at this time. Wet removal has been treated the least satisfactorily. Experimentally, washout coefficients, or a power law, have been used. A large unknown in the wet removal process is the SO_2 to $SO_4^=$ incloud transformation, an issue that is beginning to be addressed by the ANL model. Scott[48] has done some interesting work on the incloud process and has devised a chemical and dynamical model to describe SO_2 transformation within clouds during winter and summer storms.

INTERNATIONAL ASPECTS OF TRANSPORT

Prevailing Meteorology

As was shown earlier in Figures 3-1, 3-2, and 3-3 and in more detail in Figure 3-23, winds from the southwest occur with great frequency over the eastern United States, on the whole, favoring transport of pollutants from the United States to Canada. However, in a study of transport at the New York-Ontario border using data from four nearby upper-air meteorological stations, Neiman[49] found that a frequent Canada-to-United States flux occurs, with winds favoring transport from highly industrialized Ontario towards upstate New York and New England occurring nearly 50 percent of the time. Obviously, the meteorology in the eastern United States and Canada makes regional transport an international problem.

Two meteorological weather patterns that occur regularly are particularly favorable for long-range transport and subsequent dry or wet deposition. In the winter, strong winds from the northwest and southwest transport emissions rapidly from urban areas to rural parts of the region, and in the summer, stagnant high-pressure systems can persist over the east for periods of a week, promoting air pollution buildup over the region.

Transboundary Flux Estimates

The pollutant flux across the United States-Canadian border can be estimated by either correlating available air quality and meteorological data or

Atmospheric Transport, Transformation and Deposition Processes   123

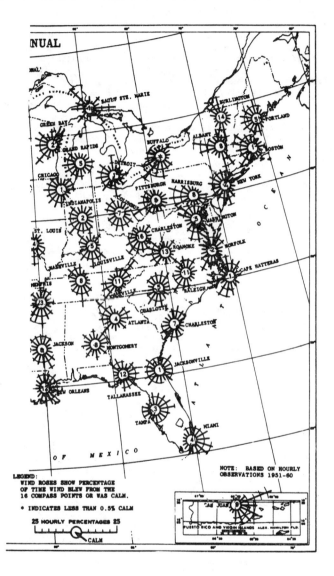

Figure 3-23. Mean annual surface wind roses in the eastern United States.[1]

using the transport and deposition models. Some initial work has been done with air quality data. For example, Fleming[50] used the Ontario Hydro network to show that high background SO_2 in Ontario, observed primarily in the winter, coincided with winds favoring transport from the United States. At this time, however, the primary source of transboundary flux estimates has been from models.

Preliminary estimates of the transboundary flux by the ASTRAP model, described earlier in Section 3, indicate that the United States contributes 4 to 5 times as much sulfur to Canada as it receives. As would be expected from the seasonal wind patterns, the summer flux is greater than the winter flux. Galloway and Whelpdale[21] estimate the difference between the United States-to-Canada and Canada-to-United States sulfur flux to be approximately a factor of three. The actual annual flux estimates are shown in Table 3-8. In addition, seasonal transport estimates based on the EURMAP-1 model indicate that the ratio of United States-Canada transport to Canada-United States transport is 1.3 in the winter and 3.2 in the summer.

TABLE 3-8. TRANSBOUNDARY FLUX ESTIMATES,[a] U.S.-CANADA RESEARCH CONSULTATION GROUP[51] (Tg S y^{-1}, MILLIONS OF METRIC TONS OF S PER YEAR)

	Method of estimation	Canada to USA flux	USA to Canada flux
I	Statistical trajectory model (ASTRAP) (Shannon, 1979)	0.5	2.
II	Simple advection and decay model (Galloway and Whelpdale, 1979)	0.7	2.

[a] Based on emissions east of approximately 92°W.

Insight into transboundary transport can be gained from estimates of the contribution from sources in the United States and Canada to each other's total sulfur deposition. In Table 3-9, EURMAP-1 and ASTRAP estimates of the amount of sulfur deposited in eastern United States and Canada are broken down into incremental contributions by United States and Canadian sources.

Both models indicate that Canadian source emissions contribute less than 5 percent of the total sulfur deposited in the United States. However, Canadian sources made a significant contribution in northern New York and northern New England. The model results also indicate that United States sources contribute about the same portion of total sulfur deposited in Canada as do Canadian sources. The Galloway and Whelpdale[21] deposition estimates based on measurements are included for comparison. The model total deposition estimates for Canada are well below the observed values, and the EURMAP-1 estimates for United States deposition are a factor of two greater than the Galloway and Whelpdale estimates.

TABLE 3-9. TOTAL SULFUR DEPOSITION ESTIMATES (MILLION METRIC TONS)

			Contribution eastern U.S. sources	From eastern Canada sources	Total
EURMAP-1	Eastern U.S.	wet	-	-	-
		dry	-	-	-
		total	10.8	0.4	11.2
	Eastern Canada	wet	-	-	-
		dry	-	-	-
		total	0.7	1.2	1.9
ASTRAP (2 summer months)	Eastern U.S.	wet	0.554	0.019	0.573
		dry	0.455	0.017	0.472
		total	1.009	0.036	1.045
	Eastern Canada	wet	0.132	0.109	0.241
		dry	0.048	0.064	0.112
		total	0.180	0.173	0.353
Galloway and Whelpdale[21] based on measurements	Eastern U.S.	wet	-	-	2.5
		dry	-	-	3.3
		total	-	-	5.8
	Eastern Canada	wet	-	-	3.0
		dry	-	-	1.2
		total	-	-	4.2

International Research

Mathematical models that are capable of estimating the transboundary transports have been developed by Battelle, Pacific Northwest Laboratory (PNL), Argonne National Laboratory (ASTRAP), and SRI International (EURMAP-1). In Canada, the Federal Atmospheric Environment Service and the Ontario Ministry of the Environment have large-scale models under development. The models are continually being refined to improve their treatment of the transport, transformation, and removal processes.

In addition to modeling studies, available air quality and meteorological data are starting to be used to analyze the transboundary transport processes. Ontario Hydro has set up an extensive SO_2 network that they are using to study statistical trends and correlations between air quality and wind direction. In the United States, the EPA is funding research efforts using available meteorological data to perform trajectory analyses.

Global Component of Acid Rain

The pH of precipitation is being measured at a number of remote sites as part of monitoring programs being carried out by the National Oceanic and Atmospheric Administration and the World Meteorological Organization. These measurements will make it possible to investigate the existence and nature of worldwide background pH levels, and, over time, provide a basis for estimating global trends in precipitation acidity.

Early results, listed in Table 3-10, show considerable variability from site to site and from year to year, and a great deal of variation about the mean at each site, as evidenced by the ranges presented in the table. The data reported in Table 3-10, when averaged over the seven sites, yields a mean pH of 5.79. The average range in pH for the various sets of data covers 2.49 pH units. The greatest range, from 2.60 to 7.10, was observed at Glacier Park, Montana, during the 1972 to 1976 period.

The difficulty in trying to determine a background pH is further demonstrated in a study of the local variation found on the island of Hawaii reported by Miller.[51d] He found a significant increase in acidity between a sea level site at the east end of the island and a site near the top of the Mauna Loa volcano. These differences remain unexplained, but volcanic outgassing is among the suggested causes.

TABLE 3-10. AVERAGE VALUE OF pH FROM REMOTE NOAA/WMO SITES[51a-51c]

Location	pH	Range	Years	No. of values
Mauna Loa, Hawaii	5.30	3.84-6.69	1973-1976	28
Pago Pago, Samoa	5.72	4.74-7.44	1973-1976	30
Prince Christian Sound, Greenland	5.73	4.70-6.81	1974-1976	13
Valentia Island, Ireland	5.43	4.2-6.8	1973-1976	48
Glacier Park, Montana	5.78	2.60-7.10	1972-1976	33
Pendleton, Oregon	5.90	4.67-7.60	1972-1976	35
Ft. Simpson, NWT, Canada	6.27	5.22-7.14	1974-1976	14
Mauna Loa, Hawaii	5.84	4.76-6.60	1976-1978	24
Pago Pago, Samoa	6.00	5.55-6.51	1977-1978	12

INTERREGIONAL DIFFERENCES

Measurement of Acidic Precipitation

Background--
As pointed out in Section 1, researchers have noticed that the pH of rainwater in many areas is lower than would be expected for a system of carbon dioxide dissolved in pure water. Acidic precipitation was first noticed in Europe in the 1950s, particularly in the Scandinavian countries of Norway and Sweden. Since then, the European Atmospheric Chemistry Network has been set up to collect data, and trends toward increasing acidity have now been documented.[7]

Concern over acidic precipitation spread to the United States in the early to mid 1970s. Likens and Bormann[52] noted the high acidity of New Hampshire lakes, particularly those of the Hubbard Brook watershed, a 75,000-acre experimental forest. Because no direct pH measurements had been taken in that area before the 1960s, no firm trend could be established. However, they made the supposition that precipitation had been much less acidic before 1930 based on the fact that the concentration of carbonate measured in rain at that time was higher than that which would be expected if strong acids were also present. Because the problem in Europe has been linked to long-range transport of air pollutants, they suggested that part of the problem could be emissions from midwestern stationary sources. Since that time, acidic precipitation has also been documented throughout the northeastern United States,[53,54] in northern Minnesota,[55] near the Continental Divide in Colorado,[56] in the Los Angeles basin,[57] in areas of the Northwest,[58] in Florida,[59] and in areas of Canada. The areal extent of precipitation acidity in North American is illustrated by the pH contours for the United States and Canada in Figure 3-24.

Long-Term Trends--
Because of the lack of long-term monitoring data, acidity trends in the United States have been poorly defined and necessarily based on calculated pH. Cogbill and Likens[53] pieced together precipitation data from the 1950s and 1960s and calculated the pH by assuming a stoichiometric relationship between the major chemical ions in rainwater. By comparing predicted pH values with actual measurements taken in the early 1970s, they estimated an error of 0.1 pH units. Maps illustrating their work appear in Figures 3-25 and 3-26. In the absence of more reliable data, their study has often been cited in estimating long-term trends. It shows both an increase in the area affected and lower pHs for precipitation appearing in New England and New York. However, the validity of their conclusions has been questioned for the following reasons.

1. Sampling methods before 1972 did not associate acid precipitation with an event and collected total wet plus dry deposition.[61]

2. Precipitation samples were not stored correctly from the earlier time period[61] and it has been noted that measurements taken in the field often differ from laboratory measurements of the same sample.[62]

128   Acid Rain Information Book

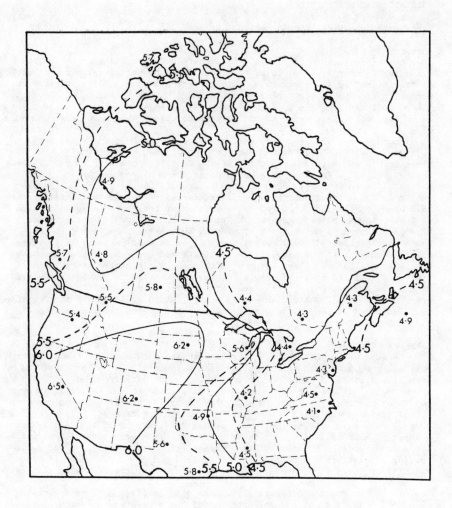

Figure 3-24. Precipitation amount weighted mean pH at North American WMO stations for 1974-1975.[60]

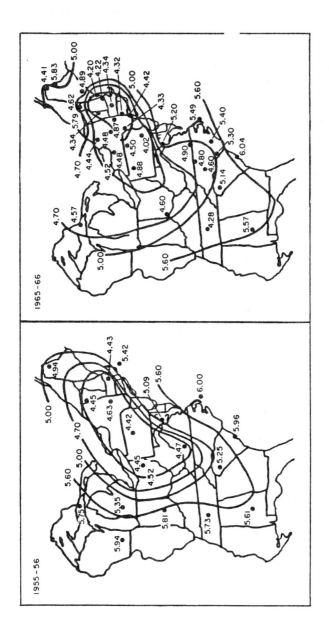

Figure 3-25. The weighted annual average pH of precipitation in the eastern United States in 1955-56 and 1965-66. (Modified from Likens et al.).[53]

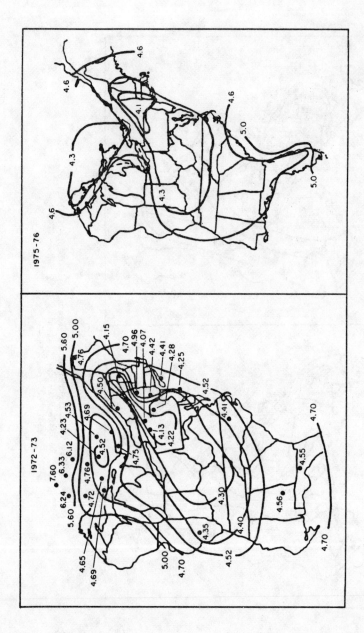

Figure 3-26. The weighted annual average pH of precipitation in eastern North America in 1972-1973 and 1975-1976 (modified from Likens et al.).[53,28]

3. Calculated values have been assigned a margin of error of 0.5 pH units by one critic.[63]

4. The sampling sites for the different time periods were not the same.

Because pH has been found to vary between locations and storms, some critics have reanalyzed the data, examining trends at individual sites.[61,63,64] The initial comparisons between 1955-56 and 1965-66 published by Cogbill and Likens[53] had 10 common sites. Four of those showed increases in pH, two showed decreases, and the other four remained the same. Only two of the same sites were measured in the 1955-56/1972-73 comparison, one showing increasing pH, and the other showing a decrease. Between 1965-66 and 1972-73, eight stations were common, with pH increases at three, decreases at two, and no changes at three. It has been stated that if only the common stations were compared, no trend could be deduced.[61,63,64]

In August of 1980, the Utility Air Regulatory Group (UARG) requested Environmental Research and Technology, Inc. (ERT) to examine further the quality and analysis of historical data regarding precipitation chemistry.[81] The data which were collected in the eastern United States were examined for the consistency of sampling and chemical analysis methods used. Attempts were made to quantify uncertainties, biases or equivalence associated with (a) sampling, analytical and calculational methods to deduce precipitation acidity, (b) use of data from different times and conditions, (c) failure to collect brief or light rainfall, (d) failure to collect initial precipitation in an event, (e) failure to analyze for magnesium in certain samples, and (f) failure of previous investigators to consider all available data. The results of these assessments were examined in conjunction with interpretation of spatial and temporal changes showing apparent trends in increased precipitation acidity over the eastern United States, taking into account natural variability in pH, climatological influences and the role of cation contributions associated with dust scavenging. ERT has stated that when the data from the few midwestern monitoring sites are interpreted on the basis of an 0.5 pH measurement uncertainty, no evidence of westward expansion of acidity can be established.[63] ERT also has pointed out that the apparent increase of acidity in the southeast shown by published isopleth maps (Figures 3-25 and 3-26) may well have resulted from the use of an incomplete data base.[63] Cogbill and Likens themselves pointed out that the decrease in pH in New York and New England evident in the 1965-66 data was primarily caused by increased resolution from additional monitoring sites.[53] The supposition of a trend toward increasing acidity has also been disputed on the grounds that the nine sites monitored by the United States Geological Survey from 1965 to 1978 in New York State where acid precipitation does occur have shown no significant trend over the time period. Time series of these sites show rather large variations over the entire sampling period as seen in Figure 3-27. Even the monthly data from Hubbard Brook showed no strong evidence of a trend toward increasing acidity.[64,65] It has also been suggested that the apparent trend toward more acidic rain may be caused by factors other than the increase of strong acids in precipitation. Stensland suggests that rather than an addition of anions, (such as $SO_4^=$

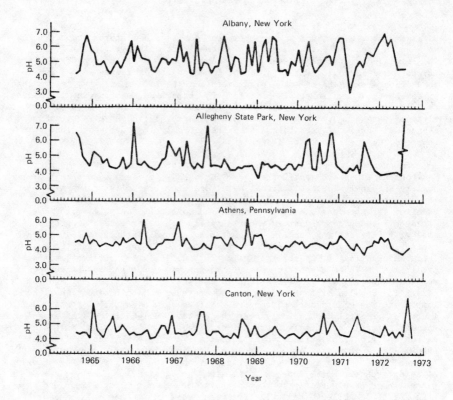

Figure 3-27. History of acidic precipitation at various sites in and adjacent to the State of New York.

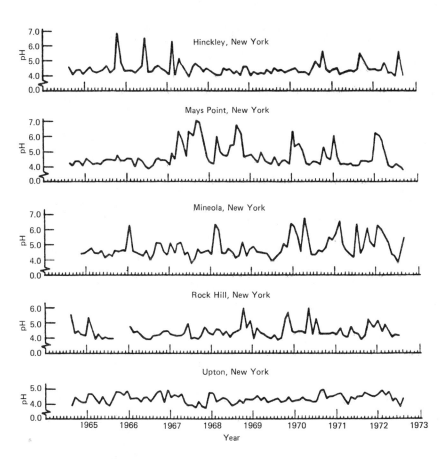

Figure 3-27 (continued).

and NO_3^-) there may have been a decrease in cations (such as Ca^{++} and Mg^{++}) from scavenged soil dust.[66] A drought occurred in the mid-1950s followed by a less severe drought in the mid-1960s with a wet period occurring during the later measurements.

The eastern United States is not the only area of the country where acid precipitation has been measured and trends proposed. Liljestrand and Morgan[57] found a volume-weighted, mean pH of 4.06 from measurements in the Pasadena, California area in 1978. This is a drop from calculated pH levels for before 1970, which were mostly above 5.6. It was noted, however, that the 1978 measurements were taken during anomalous weather conditions; data were collected while California was in a drought and two tropical storms accounted for 20 percent of the rain. Lewis and Grant[56] documented acidic rain in the Como Creek watershed near the Continental Divide in Colorado. Within widely scattered data they found a significant downward trend in pH for the three years of data collection, with the regression pH dropping from 5.43 to 4.63, but noted that such a trend should not be extrapolated. No previous study documented pH for that area. Measurements in Florida showed that the northern three-quarters of the state receives precipitation with an annual average pH below 4.7 as seen in Figure 3-28a.[59] The countours tend to hug the coast presumably because of the neutralizing effect of the maritime air. No long-term pH data had been collected in Florida, but calculated values for five locations indicated that the pH was greater than 5.6 in the mid-1950s. Half the observations taken along the Georgia coast showed pH levels less than the acid-defining value of 5.6.[67] A two-year study in the New York Metropolitan area showed a mean pH of 4.28.

North American trends have also been proposed by looking at the pH records stored in glaciers and continental ice sheets. Present-day values tend to be about pH 5, but Greenland ice from 180 years ago showed pH values ranging from 6 to 7.6.[28]

Short-term or Seasonal Trends--
Several studies have shown a seasonality in the precipitation data. After studying data from the northeast, Likens and Bormann[52] noted that in general, summer rains are more acidic than winter precipitation. Other studies have noticed the same thing. Precipitation in Florida is more acidic in summer than in winter by 0.2 to 0.3 pH units,[59] as demonstrated in Figure 3-28b. It was supposed that convective showers of summer are more efficient at scavenging sulfates and nitrates from the atmosphere than are winter frontal storms. It was noted that 65 percent of the precipitation fell during the summer months. Sulfates showed similar temporal trends as acidity but nitrates did not show seasonal variations. Studies in Georgia[67] and the New York Metropolitan area[54] also showed lower pH values in the summer. The lower summertime pH levels in Metropolitan New York were associated with air mass showers and thunder showers as well as higher sulfate levels.

Current Monitoring--
The historic precipitation data have been sparse and not readily comparable. Recent concern over the possible effects of acidic precipitation

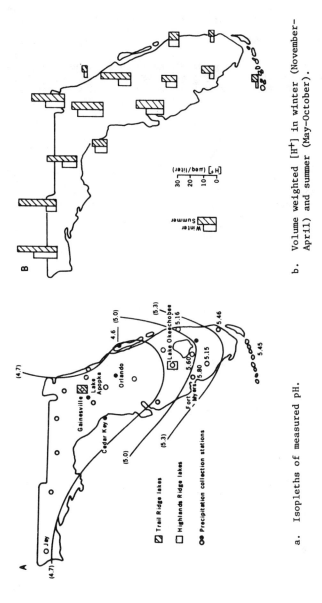

Figure 3-28. Patterns of acidity in Florida precipitation (1978-1979).[59]

a. Isopleths of measured pH.

b. Volume weighted [H$^+$] in winter (November-April) and summer (May-October).

has changed this. The World Meteorological Organization has 15 monitoring stations in the United States; the National Atmospheric Deposition Project Network monitors more than 50 stations in the United States for both wet and dry precipitation; and MAP3S runs eight stations.[68] The Electric Power Research Institute (EPRI) collects data from nine sites. Other smaller networks have also been set up. In spite of past problems resolving differences in data collection, one recent study shows that good agreement may be attained when some of the present networks are compared.[62] In this study, the data from EPRI and MAP3S sites combined well for H^+, NO_3^-, SO_4^{2-}, and NH_4^+, although differences were noticed for Cl^- and Na^+.

Contributing Components to Acidity--

Most of the studies on acidic precipitation have shown that the acidity is mainly caused by strong acids, especially sulfuric acid (H_2SO_4) and nitric acid (HNO_3). A few investigators, however, have suggested that weak acids may play an important role in contributing to the acidity of rainfall.[67,69] Most of the recent data attribute 60 to 70 percent of the acidity to sulfates, 30 to 40 percent to nitrates, and the remainder primarily to chlorides with a host of other contributing acids in small amounts for the northeastern United States.[28,65,70,71,72] The proportion of nitrates is increasing.[52] In spite of the fact that the sea salt contributions stayed the same, Cogbill and Likens noted the proportional changes below.[52]

	Sulfates	Nitrates	Chlorides
Predicted 1955-56	78%	22%	0
1973	65%	30%	5%

Measurements from Florida show that the ratio of sulfuric acid to nitric acid ranges from 2.0:1 to 2.5:1 with little fluctuation in nitrates but fluctuations in sulfates similar to that of the pH, both spatially and with time.[59] The ratio of sulfuric to nitric acid west of the Mississippi is about 1:1 to 1:1.4.[65] At Pasadena, California, the ratio is about 1.33:1.[57] The major component at Como Creek, Colorado, has been found to be nitric acid, although sulfates are also important.[56]

Monitoring of Lakes--

Acidic measurements of lakes and other water bodies have played an important role in the literature citing the role of acidic rainfall. The increasing acidity of lakes has been associated with acidic precipitation, and the distribution of acid lakes seems to be congruent with the distribution of acid precipitation.[28]

In Florida, the 12 lakes measured in areas of acidic rain (the Trail Ridge Lakes) now show a pH 0.3 to 0.9 units lower than in the late 1950s and 1960s. For example, Lake Brooklyn had an average pH of 5.5 during 1957 to 1960 which dropped to 5.0 for 1967 to 1972 and was measured at 4.9 during 1977 to 1979. In contrast, the eight lakes measured in the south-central part of the state (Highland County) where rainfall is not as acidic have shown no significant long-term trends toward acidification. All of the lakes measured are considered to be soft water lakes with a low buffering capacity. The locations are indicated on Figure 3-28.[59]

It has been noted, however, that lakes in the same area receiving the same rainfall often show different pH levels,[64] thus making it difficult to attribute the acidity solely to precipitation.

Sources of Acidic Components

Difficulties in Assessing Sources--
As explained in Section 2, no firm conclusions can be made on the exact sources of acid-producing components in precipitation. On a global basis, the relative contributions of man and nature are even controversial, although it has been estimated that the anthropogenic sources of SO_2 outweigh the natural ones by 19 to 1 in the eastern United States.[65] The ratio is not so clear for NO_x sources. Also, the processes relating to long-range transport of atmospheric components are not yet well understood, as has been discussed earlier in Section 3. Determination of sources requires an understanding of the chemical conversions and meteorological complexities of air mass movement. Beyond a distance of 150 to 200 miles, the transported pollutants can rarely be distinguished from the background concentrations. However, if the problem of acidic precipitation is to be fully evaluated, attempts must at least be made to indicate the most likely sources. Most evaluations begin by assuming long-range transport from stationary sources of SO_2. Methods of attributing the measured acidity to sources include association with fuel use patterns, evaluation of the $SO_2:NO_x$ ratio, analysis of trajectories, and results of modeling, as discussed below.

Nonquantitative Methods--
One of the simplest methods of attributing acidic precipitation measurements to sources is merely associating them with emission patterns. Likens and Bormann noted a correlation between the precipitation pH in New York State and the changes in SO_2 and NO_x emissions resulting from fuel use patterns in their 1974 study.[52] Since that time, similar correlations have been noticed in other areas. A Florida study[59] suggests that the increasing acidity of rain is directly related to the industrialization of the area and the resulting increase in SO_2 and NO_x emissions. The higher pH levels along the coast were attributed to the neutralizing effects of the marine air. The situation is slightly more complicated in the Como Creek watershed of Colorado.[56] The high acidities measured there could not be definitely attributed to increasing pollution from nearby Denver to the east because westerlies predominate in that area. The possible explanations offered were that the sporadic upslope local easterlies may be more important in pollutant transfer than the prevailing winds, or that the widespread increases in NO_x emissions throughout the West are transported in. Precipitation analyses showed that contributions from nitrates are important in that area.

It has been suggested that the relative contribution of sulfates and nitrates to the precipitation acidity may be a good indicator of pollution sources.[72] If nitrates are generally attributed to mobile sources and sulfates to stationary sources, the acidity in the West can be attributed to increasing automobile use.[72] The MAP3S data shows a correlation between sites for sulfates but wide variations in nitrates. The nitrates in precipitation are also increasing faster than sulfates. Sulfur dioxide emissions have not changed greatly since 1960, but NO_x emissions have increased by 120 percent.[71] This suggests a correlation between pH and NO_x emissions. The Pasadena study included a statistical analysis comparing the chemical properties of rainwater with those of various possible sources.[57] It was found that 35 percent of the total residue was from nitrates produced by NO_x air pollution and 20 percent from the sulfate formed from SO_2 pollutants with smaller percentages produced by soil dust, sea salt aerosol, fuel fly ash, ammonium from ammonia, automobile aerosol, and cement dust. The ratio of nitrates to sulfates in that area (1.32) compares well with the ratio of NO_x to SO_2 emissions from stationary sources in the Los Angeles basin (1.45).

Trajectory analysis is another method for determining possible sources of acidic components. Cogbill and Likens[53] looked for extra-regional sources of central New York acidity by drawing geostrophic trajectories backward in time three days using a 500-millibar (mb) windfield. Figure 3-29 shows the calculated movements and the average pH values that were observed for the different corridors. The most acidic depositions would result from air moving from the southwest over the pollution sources of the Ohio Valley. They noted that the analysis did not take into consideration the possible influence of local sources because these would be more appropriately found using an 850-mb windfield. Mixing effects were not considered. A later study of the Ithaca, New York area[73] used a computerized trajectory model developed by the National Oceanic and Atmospheric Administration's Air Resources Laboratory (ARL). Winds were averaged for a layer between 300 and 1500 m above ground, and each sector was categorized by the usual type of precipitation event for that direction. Good agreement was found between high acidity and high concentrations of $SO_4^=$, NO_3^-, and Cl^-, and once again, precipitation was most acidic when originating from areas to the southwest. The summer acidities were found to be higher than those for winter within this sector. This study has been criticized because air mass movement is much more complicated than that hypothesized within corridors.[63] A study of the New York Metropolitan area[54] separated precipitation events according to the sector within which the storm originated based on trajectory analyses of NWS data. The differences noticed in pH from each direction are indicated in Figure 3-30. Lower pH levels occurred when the storms approached from areas with high emissions to the west and southwest. However, the types of storms from these directions tended to be those which may be more efficient in scavenging pollutants. The comparisons were made with easterly winds, which had alkaline input from the sea salt.

Atmospheric Transport, Transformation and Deposition Processes   139

a. Three days prior to precipitation event in New York.

b. Two days prior to precipitation.

c. One day prior to precipitation.

Figure 3-29. Geostrophic back trajectories at the 500 mb level as analyzed by Cogbill and Likens.[53] Corridors are arbitrary.

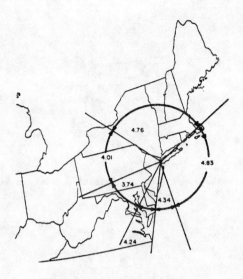

Figure 3-30. The variation of mean precipitation pH by directional sector of approach for the New York Metropolitan area.[54]

Modeling Results--
The EURMAP-1 model, reviewed earlier in Section 3, has been applied to determine interregional exchanges of sulfur using emissions sources from SURE and SAROAD data bases.[39] The results are shown in Tables 3-11 and 3-12. Both the total calculated annual averages (Table 3-11) and percent contributions (Table 3-12) are included. These may be referenced for location with Figure 3-31. These modeling results, while obviously tentative, are such as to raise questions about some of the widely held ideas regarding the sources of acid rain precursor pollutants in the eastern U.S. For instance, the tables indicate that for Region II (New York State) 28 percent of the total sulfur deposition (TSD) comes from within Region II, while 25 percent of the TSD is from southeastern Canada and New England, 13 percent is from Region V-South, and 24 percent is from Region III. For Region I (New England), 44 percent of the TSD is from within the region, 32 percent is from southeastern Canada and New York, and only 18 percent is from Regions III and V-South. For South Ontario, according to this model, 61 percent of the TSD is from southeastern Canada and 39 percent is from the eastern U.S. For South Quebec, 71 percent of the TSD is from southeastern Canada and only 29 percent is from the U.S. The model is still considered to be in the developmental stage, and the limiting assumptions have been discussed earlier. This

TABLE 3-11. INTERREGIONAL EXCHANGES OF AIRBORNE SULFUR: TOTAL CONTRIBUTIONS IN 1977 AS PREDICTED FROM EURMAP-1 MODEL

Emitter region	Total contributions to S depositions within receptor regions												
	1	2	3	4	5	6	7	8	9	10	11	12	13
1 VIII-North	10.	1.	0.	2.	0.	0.	8.	0.	0.	0.	0.	0.	0.
2 V-North	3.	655.	290.	46.	0.	3.	229.	6.	24.	78.	50.	18.	23.
3 S. Ontario	0.	66.	820.	2.	0.	1.	49.	2.	7.	74.	87.	40.	87.
4 VII	1.	43.	10.	367.	0.	26.	137.	22.	41.	12.	3.	2.	2.
5 VIII-South	0.	0.	0.	0.	0.	0.	0.	6.	0.	0.	0.	0.	0.
6 VI-East	1.	4.	1.	40.	0.	401.	7.	35.	6.	1.	0.	0.	26.
7 V-South	2.	186.	145.	135.	1.	14.	1566.	59.	425.	520.	92.	30.	2.
8 IV-South	0.	8.	7.	16.	0.	44.	31.	949.	279.	25.	2.	0.	26.
9 IV-North	0.	19.	24.	11.	0.	13.	221.	108.	929.	159.	15.	1.	6.
10 III	0.	1.	11.	57.	0.	1.	178.	14.	141.	1363.	179.	56.	14.
11 II	0.	1.	53.	0.	0.	0.	1.	1.	4.	37.	204.	65.	14.
12 I	0.	0.	1.	0.	0.	0.	1.	0.	2.	91.	8.	207.	22.
13 S. Quebec	0.	2.	105.	0.	0.	0.	1.	0.	0.	2.	8.	41.	204.
Total (Kton S)	18.	997.	1514.	621.	1.	503.	2422.	1197.	1856.	2280.	732.	467.	407.

TABLE 3-12. INTERREGIONAL EXCHANGES OF AIRBORNE SULFUR: PERCENT CONTRIBUTIONS IN 1977 AS PREDICTED FROM EURMAP-1 MODEL

Emitter region	Percent contributions to S depositions within receptor regions												
	1	2	3	4	5	6	7	8	9	10	11	12	13
1 VIII-North	55.	0.	0.	0.	0.	0.	0.	0.	0.	0.	0.	0.	0.
2 V-North	19.	66.	19.	7.	6.	1.	9.	0.	1.	3.	7.	4.	6.
3 S. Ontario	3.	7.	54.	0.	0.	0.	2.	0.	0.	3.	12.	9.	21.
4 VII	1.	4.	1.	59.	0.	5.	6.	2.	2.	1.	0.	0.	0.
5 VIII-South	0.	0.	0.	0.	92.	0.	0.	2.	0.	0.	0.	0.	0.
6 VI-East	7.	0.	0.	6.	1.	80.	0.	3.	0.	0.	0.	6.	6.
7 V-East	9.	19.	10.	22.	1.	3.	65.	5.	23.	23.	13.	0.	1.
8 V-South	1.	1.	0.	3.	0.	9.	1.	79.	15.	1.	2.	0.	1.
9 IV-South	0.	2.	2.	2.	0.	9.	9.	1.	50.	7.	0.	0.	6.
10 IV-North	2.	1.	4.	1.	0.	0.	7.	1.	8.	60.	24.	12.	1.
11 III	1.	0.	0.	0.	0.	0.	1.	0.	0.	2.	28.	14.	5.
12 II	0.	0.	4.	0.	0.	0.	0.	0.	0.	0.	12.	44.	5.
13 S. Quebec	0.	0.	7.	0.	0.	0.	0.	0.	0.	0.	1.	9.	50.

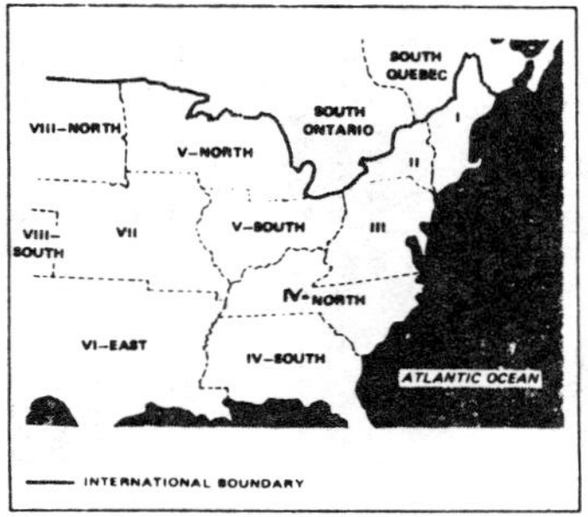

Figure 3-31. Source regions used for the EURMAP-1 model.

model appears to be the only model used so far to give results on an interregional basis, and it does not include the western portion of the United States. No similar work has yet been performed using NO_x sources.

Local Sources of Acidic Components--

Section 2 discussed the natural and anthropogenic sources of acidic components that would be subject to atmospheric transport. Most of these sources emit precursor pollutants and affect receptors after transformations have occurred within the atmosphere. Others, such as oil-fired facilities, may emit sulfates or other acid forming pollutants directly and therefore affect local receptors. When the effects of acidity are discussed, it must be remembered that natural receptors such as lakes and waterways receive acidic components from local drainage water as well. Many of the pH measurements have been taken in lakes that are subject to acids injected from ground water, bogs, polluted streams, mine drainage, and many other sources. Strong acids such as sulfuric and nitric acids are not the only ones contributing to lake acidification. Although the strong acids are important, weak acids such as humic and fulvic acids, which are produced by biodegradation of vegetable matter, and inorganic weak acids based on hydrated aluminum, iron, and species of silica, are also present in significant amounts, indicating interaction with the ground during runoff.[74] These acids may also contribute to the changes noticed. Rosenqvist[75] suggested that changes in the environment in the past several decades may have altered patterns of weak acid production significantly. Factors of possible influence include land use for agriculture, with its potential for fertilizer runoff, expansion of commercial forestry, control of forest fires, and changes in weather patterns.

How the acidified water enters a lake may also be important. Even lakes that are not acidic to begin with may be very responsive to episodic additions such as those occurring during a spring snowmelt. These episodes may result from release of acids stored in the snowpack. As the snow melts, it percolates through the snowpack and leaches stored acids.[76] Such acids may be either strong or weak, and the exact processes and contributing factors are not yet completely understood.

Cumulative Buildup of Acidic Components

Acid Buildup in Lakes--

The accumulation of acidic components over a period of time can have an effect on water bodies. Most lakes have the ability to buffer or neutralize the incoming acids. In the softwater lakes that are sensitive to additions of strong acids, the predominant anion is bicarbonate (HCO_3^-), the dissociated form of the weak carbonic acid (H_2CO_3).[77] As long as bicarbonates persist, the added acids are neutralized and the pH does not change. However, when the bicarbonate is exhausted, further addition of acid causes the pH to drop more rapidly. This point is sometimes referred to as the threshold of lake acidification.

Figure 3-32 demonstrates the response of two individual lakes at different levels of sensitivity as atmospheric sulfate loading is increased. The result of sulfate loading past the lake's threshold is demonstrated by the

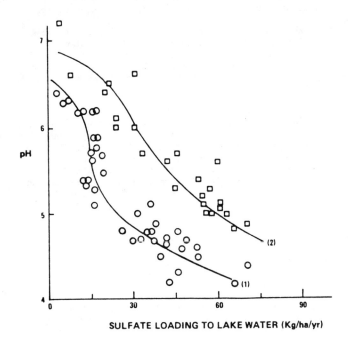

Figure 3-32. Relationship between acid loading and pH change for:[72]
  O very sensitive and
  □ moderately sensitive surroundings.

sharp drop in pH.[72,78] When the lake has passed this point, it is more likely to show large fluctuations in response to episodal additions.[77] Note, however, that these curves were developed for Scandinavian lakes considered to have a sensitivity to acid precipitation and do not necessarily reflect the response of lakes in most of the United States and Canada. Each lake has its own response curve based on its watershed characteristics.

Based on European data, an empirical relation has been developed to predict the threshold for fresh water acidification; for instance, a system with 80 microequivalents of bicarbonate could be expected to drop below pH 5 if the long-term average pH of precipitation is below 4.3.[28] This method may not be valid for widespread application, however, because various factors affect the sensitivity of areas to acidic precipitation.

Area Sensitivity to Acidic Precipitation--

Not all lakes are equally susceptible to acidification. Lakes in New England, the Adirondacks, and northern Minnesota are showing signs of stress, whereas many of those sampled in western areas of the country that also receive acid precipitation have not experienced pH changes. Even in the same area, lakes display varying pH levels. It seems that the sensitivity of a given lake depends on many interacting factors in ways not yet well understood. The factors that are generally looked at can be grouped as:

- meteorological patterns affecting precipitation,
- hydrology of the lake,
- watershed characteristics, and
- soil and bedrock geology.

The effects of meteorology are important because they determine what the precipitation contains and where it falls, as discussed previously in Section 3. Where the air comes from determines what pollutants it may contain and also the neutralizing agents that may be included, such as suspended soil particles or sea salt. The topography of the area helps determine where the acidic precipitation is dropped.[79]

The hydrology of the lake includes all the inputs and outputs of lake water as well as its neutralizing capabilities. The characteristics of stream water, ground water, and runoff from areas of high chemical reaction are important.[78]

The characteristics of the watershed area help determine sensitivity. Some of the sulfates and nitrates fall as dry deposition and are washed into the lakes during a rain shower. Decomposition of materials in the watershed may alter the ion balance. Hence, factors such as watershed area to lake volume ratio and vegetative cover are important considerations.[72] Patterns of land use may largely affect the watershed area and thus the chemical composition of the runoff,[63] as well as behavior of the organisms.[28]

Probably the most discussed determining factors when evaluating area
sensitivity are the underlying bedrock and the topsoil that covers it.
Highly sensitive areas are usually underlain by primarily silaceous bedrock
such as granite, some gneisses, quartzite, and quartz sandstone.[28] Such
rock types are highly resistant to weathering and thus cannot contribute to
the ionic balance of surface water. Areas with silaceous bedrock include
Scandinavia, the Canadian Shield, the Rockies, New England, the Adirondacks,
and smaller areas elsewhere.[28] The soil overlying the bedrock does not
necessarily have the same composition as the bedrock below, as is evident
when areas with a history of glaciation or marine transgressions are considered.
For example, Florida is underlain by bedrock that would indicate low
sensitivity, but the areas where acidic lakes have been reported have soils
composed of highly weathered material with low cation exchange capacities
that are unable to provide much buffering capacity.[59] A map of geologically
sensitive areas appears in Section 4, Figure 4-2.

Although all the factors above are considered important in determining
susceptibility to lake acidification, they are not considered to be a complete
list and the relative importance of each has not been quantified.

A large study currently underway is attempting to quantify some of the
factors mentioned by extensively studying three lakes that are within a 20-
mile radius in the Adirondacks yet show differing acidities.[80] The study
was undertaken by EPRI, and it is hoped that the outcome will be a model
quantifying the interrelationships of a lake ecosystem that determine its
acidity.

Models to Determine Acceptable Loadings of Acidic Materials

The models previously discussed in this section are expected to be useful
in converting emissions of sulfur compounds at one place to deposition in
another. However, they are only transport models and not designed to define
how much acidic deposition is tolerable in sensitive areas.

Developing such acidic tolerance models for crop damage or materials
effects will be very difficult due to interference from other pollutants or
due to direct introduction of chemicals by man, such as fertilizer application.
However, three models for relating deposition of sulfur compounds
to adverse aquatic effects have been described in a recent U.S./Canadian
report.[82]

The first model described is by Dickson, and is based on an empirical
relationship between sulfate loading to lake water (in kilograms per hectare
per year), and pH. The U.S./Canadian report interpreting Dickson's material
suggested tolerable sulfate loading between 9 and 17 kg/ha/yr, depending on
lake sensitivity. The report advised caution in applying this Swedish data
to North America.

The second aquatic tolerance model, also developed for Sweden, is by
Henriksen and involves two relationships. The first is the relationship
between between calcium ions and pH of lake water. The other relationship is

between excess calcium and magnesium ions in lakes, pH of rainwater, and excess sulfate in lake water. Using these relationships, the U.S./Canadian report tentatively concluded a precipitation sulfate concentration of about 40 ueq/l would protect most sensitive lakes. The report noted that Henriksen's equations have not been validated for North America, that the method does not consider pulse episodes such as snow melts, and that nitrates (which represent about one-third of the acidity of some eastern North American precipitation) are not included in the analysis.[82]

The third acidic tolerance model described is being developed by Thompson and others in Canada. It focuses on the depletion of positive ions, or cations, in watersheds. Hence, this model also relates calcium ions and pH of water bodies. This model is interesting because it addresses short-term effects of storm events or snow melts which are believed to have shock effects on fish reproduction and fish fry. The U.S./Canadian group cited above concluded Thompson's work indicated an annual sulfate loading not exceeding 5 to 7 kg/ha is necessary to protect the most sensitive lakes. They also concluded more work is required to evaluate Thompson's assumptions and model relationships.[82]

It should be noted that all of the above models attempt to relate sulfate deposition to critical changes in lake pH. No similar relationships for nitrogen compound deposition were found in the literature.

SUMMARY

The spatial and temporal distributions of acid precipitation in North America are strongly influenced by large-scale climatological features. Of particular importance are the prevailing wind patterns that transport pollutants from major industrial areas, and the location of preferred storm tracks. In combination, those two features make the acid rain phenomenon of special concern to the northeastern United States and the neighboring parts of Canada. In addition, precipitation is enhanced when moisture-laden air is forced to ascend topographic barriers, such as the Adirondacks.

Better understanding of the acid rain phenomenon and the development of source-receptor relationships require a more thorough knowledge of the chemical and physical processes acting between the sources of acid rain precursors and the receptors. Transformation processes are exceedingly complex and involve both homogeneous and heterogeneous reactions. The rates of transformation are highly dependent upon the composition of the polluted air and the amount of solar radiation present. Some reactions are completed very quickly while others proceed over periods of several days or weeks.

Pollutants are removed from the atmosphere during cloud growth and precipitation, and also by direct contact with the ground and vegetation as a result of atmosphere dispersion and gravitational settling. The first process is referred to as wet deposition and the second as dry deposition. Recent evidence suggests that the two processes may be of equal importance over western Europe and much of North America.

Regional transport and deposition of sulfur oxides in the northeastern United States and eastern Canada has been simulated by the Battelle, Pacific Northwest Laboratories (PNL), SRI International (EURMAP-1) and the Argonne National Laboratory (ASTRAP) models. In all three models, horizontal transport is calculated using an hourly updated windfield, assumed to be representative of the entire mixed layer. Chemical transformation and removal processes are accounted for with empirically based relationships. At the present, due to a lack of data as well as theoretical understanding, the nitrogen cycle has not been simulated.

The monthly-averaged SO_2 and $\overline{SO_4}$ concentration fields predicted by these models were generally within 30 percent of measured values in the vicinity of large point sources. However, predictions were off by a factor of two or more in rural parts of the modeled region. The model results indicate that the eastern United States contributes four to five times as much sulfur to Canada as it receives. Also, Canadian sources contribute only 5 percent of the total sulfur deposited in the eastern United States, while the United States contributes about the same portion of total sulfur deposited in eastern Canada as do Canadian sources.

The occurrence of acidic precipitation is widely documented throughout the United States. However, the frequently reported trend toward increasing acidity is largely inferred from composite data bases acquired by different sampling networks and procedures. Some data from the U.S. Geological Survey collected between 1965 and 1978 in New York show that while the precipitation is acidic, there is no significant trend towards increasing acidity over this time period. The number of monitoring sites has recently been substantially increased to meet the need for more definitive trend analyses and to better define the areal distribution of acid precipitation. Most of the recent data from the eastern United States shows that the acid component of rain consists of about 60 percent sulfates, 30 percent nitrates, and that much of the remainder is made up of chlorides. The proportion of nitrates has increased somewhat over time. In the West, the roles of sulfates and nitrates are more nearly equivalent.

It is currently impossible to pinpoint the sources of measured acidity, and most procedures used to investigate probable sources presuppose long-range transport. Methods used include association with fuel use patterns, examination of $SO_4^=$ to NO_x ratios, analysis of trajectories, and results of modeling.

Lakes are often monitored as natural collectors of precipitation. They are, however, also affected by acidic products from bogs, polluted streams, pesticides, mine drainage, and other local sources. Furthermore, the sensitivity of lakes to acidic input varies widely depending upon the neutralizing capacity of the soil and underlying bedrock, the hydrology of the lake, and the characteristics of the watershed.

REFERENCES

1. Environmental Science Services Administration. Climatic Atlas of the United States. U.S. Department of Commerce, Environmental Data Services, 1968.

2. Bryson, R. A., and F. K. Hare. World Survey of Climatology, Vol. 11, Climates of North America. Elsevier Scientific Publishing Company, New York, New York, 1974.

3. Korshover, J. Climatology of Stagnating Anticyclones East of the Rocky Mountains, 1936-1965. P.H.S. Publication No. 999-AP-34, U.S. Department of Health, Education, and Welfare, National Center for Air Pollution Control, Cincinnati, Ohio, 1967.

4. Holzworth, G. C. A Study of Air Pollution Potential for the Western United States. J. Appl. Meteorol., 1:366, 1962.

5. Cox, R. A. Particle Formation from Homogeneous Reactions of Sulfur Dioxide and Nitrogen Dioxide. Tellus, 26:235, 1974.

6. Committee on Sulfur Oxides. Sulfur Oxides. National Research Council, National Academy of Sciences, Washington, D.C., 1978.

7. Barnes, R. A. The Long Range Transport of Air Pollution - A Review of European Experience. JAPCA, 29(12):1219-1235, 1979.

8. Friend, J. P. A Review of the Atmospheric Chemistry Related to Precipitation. In: Proceedings, Advisory Workshop to Identify Research Needs on the Formation of Acid Precipitation, Alta, Utah, Electric Power Research Institute, Palo Alto, California, 1978.

9. Husar, R. B., D. E. Patterson, J. D. Husar, N. V. Gillani, and W. E. Wilson, Jr. Sulfur Budget of a Power Plant Plume. Atmos. Environ., 12(1-3): 549-568, 1978.

9a. Hegg, D. A. and P. V. Hobbs. Measurement of Gas-to-Particle Conversion in the Plumes from Five Coal-Fired Electric Power Plants. Atmos. Environ., 14(1): 99-116, 1980.

9b. Roberts, D. B. and D. J. Williams. The Kinetics of Oxidation of Sulfur Dioxide Within the Plume from a Sulphide Smelter in a Remote Region. Atmos. Environ., 13(11): 1485-1498, 1979.

10. OECD Programme on Long Range Transport of Air Pollutants, Measurements and Findings, Paris: Organization for Economic Cooperation and Development, 1977.

11. The National Research Council. Nitrogen Oxides. National Academy of Sciences, Washington, D.C., 1977.

12. Haagen-Smit, A. J., and L. G. Wayne. Atmospheric Reactions and Scavenging Processes. In: Air Pollution, 3rd Edition, Vol. 1. A. C. Stern, ed., 1976.

13. Marsh, A. R. W. Sulfur and Nitrogen Contributions to the Acidity of Rain. Atmos. Environ., 12(1-3):401-406, 1978.

14. Hobbs, P. V. A Reassessment of the Mechanisms Responsible for the Sulfur Content of Acid Rain. In: Proceedings, Advisory Workshop to Identify Research Needs on the Formation of Acid Precipitation, Alta, Utah. Electric Power Research Institute, Palo Alto, California, 1978.

15. Pack, D. H. Acid Precipitation - The Physical Systems. In: Proceedings, Advisory Workshop to Identify Research Needs on the Formation of Acid Precipitation, Alta, Utah. Electric Power Research Institute, Palo Alto, California, 1978.

16. Scott, W. D., and P. V. Hobbs. The Formation of Sulfate in Water Droplets. J. of Atmos. Sci., 24:54-57, 1967.

17. MacCracken, M. C. Simulation of Regional Precipitation Chemistry. In: Proceedings, Advisory Workshop to Identify Research Needs on the Formation of Acid Precipitation, Alta, Utah. Electric Power Research Institute, Palo Alto, California, 1978.

18. Fisher, B. E. A. The Calculation of Long Term Sulfur Deposition in Europe. Atmos. Environ., 12(1-3):489-501, 1978.

19. Sulfur in the Atmosphere. In: Proceedings of the International Symposium. Held in Dubrovnik, Yugoslavia, Sept. 1977. Atmos. Environ., 12(1-3), 1978.

20. Shannon, J. D. A Model of Regional Long-Term Average Sulfur Atmospheric Pollution, Surface Removal, and Net Horizontal Flux. Atmospheric Physics Section, Radiological and Environmental Research Division, Argonne National Laboratory, Argonne, Illinois.

21. Galloway, J. N., and D. M. Whelpdale. An Atmospheric Sulfur Budget for North America. Atmos. Environ., 14(14):409-417, 1980.

22. Wesely, M. L., B. B. Hicks, W. P. Dannevik, S. Frisella, and R. B. Husar. An Eddy-Correlation Measurement of the Particulate Deposition from the Atmosphere. Atmos. Environ., 11(6):561-563, 1977.

23. Argonne National Laboratory. Report to the MAP3S Precipitation Chemistry Meeting. Ithaca, New York, 1978.

24. Semonin, R. G. The Variability of pH in Convective Storms. In: Proceedings of the First International Symposium on Acid Precipitation and the Forest Ecosystem. Columbus, Ohio, 1975.

25. Hendry, C. Chemical Composition of Rainfall at Gainesville, Florida. M.S. Thesis, Department of Environmental Engineering Sciences, University of Florida, Gainesville, Florida, 1977.

26. Seymour, M. D., S. A. Schubert, J. W. Clayton, Jr., and Q. Fernando. Variations in the Acid Content of Rain Water in the Course of a Single Precipitation. Water, Air, and Soil Pollution, 10:147-161, 1978.

27. Raynor, G. C. Meteorological and Chemical Relationships from Sequential Precipitation Samples. In: Control of Emissions from Stationary Combustion Sources: Pollutant Detection and Behavior in the Atmosphere, W. Licht, A. J. Engel, and S. M. Slater, eds. Amer. Inst. of Chem. Eng. Symp. Series No. 188, Vol. 75, AIChE, New York, New York, 1979. pp. 269-273.

28. Likens, G. E., R. F. Wright, J. N. Galloway, and T. J. Butler. Acid Rain. Scientific American, 241(4):43-51, 1979.

29. Powell, D. C., D. J. McNaughton, L. L. Wendell, and R. L. Drake. A Variable Trajectory Model for Regional Assessments of Air Pollution from Sulfur Compounds. PNL-2734, Battelle, Pacific Northwest Laboratory, Richland, Washington, 1978.

30. Scott, B. C. Parameterization of Sulfate Removal by Precipitation. J. App. Met., 17:1375-1389, 1978.

31. SURE/MAP3S Sulfur Oxide Observations with Long Term Regional Model Predictions. Atm. Environ., 14:55-63, 1980.

32. Scott, B. C., and N. S. Laulainen. On the Concentration of Sulfate in Precipitation. J. Appl. Met., 18:138-147, 1979.

33. Dana, M. T., J. M. Hales, and M. A. Wolf. Rain Scavenging of SO_2 and Sulfate from Power Plant Plume. J. Geophy. Res., 80:4119-4129, 1975.

34. U.S. Environmental Protection Agency. 1973 National Emissions Report. EPA-450/2-75-007, 1976.

35. U.S. Federal Power Commission. Steam Electric Plant Air and Water Quality Control Data, 1973. FPC-5-253, 1976.

36. McNaughton, D. J., and B. C. Scott. Modeling Evidence of Inclond Transformation of Sulfur Dioxide to Sulfate. APCA Journal, 30(3):272-273, 1980.

37. Dana, M. T. The MAP3S Precipitation Chemistry Network: Second Periodic Summary Report (July 1977-June 1978). PNL 2829, Battelle Pacific Northwest Laboratory, Richland, Washington, 1979.

38. Johnson, W. B., D. E. Wolf, and R. L. Mancuso. Long Term Regional Patterns and Transfrontier Exchanges of Airborne Sulfur Pollution in Europe. Atm. Environ., 12:511-527, 1978.

39. Bhumralkar, C. M., W. B. Johnson, R. L. Mancuso, R. A. Thuiller, D. E. Wolf, and K. C. Nitz. Interregional Exchanges of Airborne Sulfur Pollution and Deposition in Eastern North America. In: Second Joint Conference on Applications of Air Pollution Meteorology, American Meteorological Society, 1980. pp. 225-231.

40. Sheih, C. M. Application of a Statistical Trajectory Model to the Simulation of Sulfur Pollution Over Northeastern United States. Atmos. Environ., 11:173-178, 1977.

41. Durst, C. S., A. F. Crossley, and N. E. Davies. Horizontal Diffusion in the Atmosphere as Determined by Geostrophic Trajectories. J. Fluid Mech., 6:401-422, 1959.

42. Huser, R. B., D. E. Patterson, J. D. Husar, N. V. Gillani, and W. E. Wilson. Sulfur Budget of a Power Plant Plume. Atmos. Environ., 12:549-568, 1978.

43. Wesely, M. L., and B. B. Hicks. Some Factors that Affect the Deposition Rates of Sulfur Dioxide and Similar Gases on Vegetation. J. Air Pollut. Control Assoc., 27:1110-1116, 1977.

44. Wesely, M. L., B. B. Hicks, W. P. Dannevik, S. Frisella, and R. B. Husar. An Eddy-Correlation Measurement of Particulate Deposition from the Atmosphere. Atm. Environ., 11:561-563, 1977.

45. Hicks, B. B. An Evaluation of Precipitation Scavenging Rates of Background Aerosol. J. Appl. Meteorol., 17:161-165.

46. Shannon, J. D. A Model of Regional Long-Term Average Sulfur Atmospheric Pollution, Surface Removal, and Net Horizontal Flux. Atmospheric Physics Section, Radiological and Environmental Research Division, Argonne National Laboratory, Argonne, Illinois, 1980.

47. Europe and United States. Naval Post Graduate School Tech. Rep. NPS 63-78004, 1978. 107 pp.

48. Scott, B. C. Predictions of In-Cloud Conversion Rates of SO_2 to SO_4 Based Upon a Simple Chemical and Dynamical Model. In: Second Joint Conference on Applications of Air Pollution Meteorology, American Meteorological Society, 1980. pp. 389-396.

49. Niemann, B. L. An Initial Evaluation of Transboundary Transport of Pollutants and Wet Deposition Between the Eastern United States and Canada. In: Session 54, Meteorology 11, Long Range Transport and Transformation of Pollutants, Air Pollution Control Association, 1980. p. 80-54.1.

50. Fleming, R. A. Long Range Transport of Sulfur Dioxide. In: Session 54, Meteorology 11, Long Range Transport and Transformation of Pollutants, Air Pollution Control Association, 1980. p. 80-54.1.

51. United States - Canada Research Consultation Group on the Long-Range Transport of Air Pollutants. The LRTAP Problem in North America: A Preliminary Overview. 1980.

51a. Environmental Data Service 1973, 1974. Atmospheric Techniques and Precipitation Chemistry Data for the World. National Climatic Center, Asheville, NC.

51b. Environmental Data Service 1975-1976. Global Monitoring of the Environment for Selected Atmospheric Constituents. National Climatic Center, Asheville, NC.

51c. National Oceanic and Atmospheric Administration, U.S. Dept. of Commerce. Geophysical Monitoring for Climatic Change No. 7. Summary Report 1978 (1979).

51d. Miller, J. P. The Acidity of Hawaiian Precipitation as Evidence of Long Range Transport of Pollutants. Publ. No. 538, World Meteorol. Org., Geneva, Switzerald, 1979. p. 231.

52. Likens, G. E., and F. H. Bormann. Acid Rain: A Serious Regional Environmental Problem. Science, V. 184, 1974.

53. Cogbill, C. V., and G. E. Likens. Acid Precipitation in the Northeastern United States. Water Resources Research, 10(6), 1974.

54. Wolff, G. T., P. J. Lioy, H. Golub, and J. S. Hawkins. Acid Precipitation in the New York Metropolitan Area: Its Relationship to Meteorological Factors. Environmental Science & Technology, 13(2).

55. Schofield, C. L. Acid Precipitation: Effects on Fish. Ambio., 5(5-6):228-230, 1976.

56. Lewis, W. M., and M. C. Grant. Acid Precipitation in the Western United States. Science, 207, 1980.

57. Liljestrand, H. M., and J. J. Morgan. Chemical Composition of Acid Precipitation in Pasadena, California. Environmental Science & Technology, 12(12), 1978.

58. Pack, D. H., G. J. Ferber, J. L. Heffter, K. Telegadas, J. K. Angell, W. H. Hockes, and L. Machta. Meteorology of Long-Range Transport. Atmos. Envi., 12:425-444, 1978.

59. Brezonik, P. L., E. S. Edgerton, and C. D. Hendry. Acid Precipitation and Sulfate Deposition in Florida. Science, 208, 1980.

60. Ecological Effects of Acid Precipitation. In: Workshop Proceedings. Prepared for Electric Power Research Institute, 1979.

61. Comments of Amax Inc. on External Review. Draft No. 1, July 31, 1980.

62. Pack, D. J. Precipitation Chemistry Patterns: A Two-Network Data set. Science, 208, 1980.

63. Environmental Research & Technology, Inc. Comments on External Review Draft No. 1 of the Air Quality Criteria for Particulate Matter and Sulfur Oxides. 1980.

64. Perhac, R. M. Testimony to the Subcommittee on Environmental Pollution of the Senate Committee on Environment and Public Works. Presented by the Electric Power Research Institute, March 19, 1980.

65. Acid Rain Coordination Committee. The Federal Acid Rain Assessment Plan. Executive Office of the President, Council on Environmental Quality.

66. Stensland, G. J. Precipitation Chemistry Trends in the Northeastern United States. In: Polluted Rain, T. Toribara, et al., eds. Pergamon Press, New York, New York, 1980. p. 87.

67. Haines, E. B. Nitrogen Content and Acidity of Rain on the Georgia Coast. Water Resources Bulletin, Vol. 12 No. 6.

68. Office of Research and Development. Research Summary, Acid Rain. EPA-600/8-79-028. U.S. Environmental Protection Agency, 1979.

69. Frohlger, J. O., and R. Kane. Precipitation: Its Acidic Nature. Science, 179, 1975.

70. McBean, G. A. Long-Range Transport and Deposition of Acidic Substances Across National Boundaries. In: 73rd Annual Meeting of Air Pollution Control Association, Montreal, Quebec, 1980.

71. Kane R. Acid Precipitation. U.S. Department of Energy, 1979.

72. Glass, N. R., G. E. Glass, and P. J. Rennie. Environmental Effects of Acid Precipitation. EPA Decision Series - Energy/Environ. June 7-8, 1979. (4th National Conference on Interagency Energy/Env. R&D Program.)

73. Miller, S. M., J. N. Galloway, and G. E. Likens. Origin of Air Masses Producing Acid Precipitation at Ithaca, New York. Geophysical Research Letters, 5(9), 1978.

74. Gloves, G. M., and A. H. Webb. Weak and Strong Acids in the Surface Waters of the Tovdal Region in S. Norway. Water Res., 13:781-783, 1979.

75. Rosenqvist, I. T. A Contribution Towards Analysis of Buffer Properties of Geological Materials Against Strong Acids in Precipitation Water. Norwegian General Sci. Res. Council. Council for Res. and Natural Sci., Oslo, Norway, 1976. p. 99.

76. Schofield, C. L. Acid Precipitation's Destructive Effects on Fish in the Adirondacks. N.Y. Food Life Sci. Q., 10(13):12-15, 1977.

77. Wright, R. F., and E. T. Gjessing. Changes in the Chemical Composition of Lakes. Ambio., 5:219-223, 1976.

78. Glass, N. R., G. E. Glass, and P. J. Rennie. Effects of Acid Precipitation. Environmental Science and Technology, 13(11), 1979.

79. Hendry, G. R., and P. W. Lipfert. Acid Precipitation and the Aquatic Environment. Presented by Brookhaven National Laboratory to the Committee on Energy and Natural Resources, United States Senate. May 28, 1980.

80. EPRI. Lake Acidification Study - Technical Work Statement. Study for the Electric Power Research Institute.

81. Hansen, D. A. and G. M. Hidy. Examination of the Basis For Trend Interpretation of Historical Rain Chemistry in the Eastern United States, Draft, November 1980.

82. U.S./Canadian Transboundary Air Pollution Working Group, Interim Report of Work Group I, Chapter 3, pp. 113 - 136, January 16, 1981.

Section 4

Adverse and Beneficial Effects of Acid Precipitation

INTRODUCTION

As discussed in detail in earlier sections of this book, recent studies have reported a pH of 4.0 - 5.0 for rain and snow in the United States, particularly in the Northeast.[1-11] It has been suggested that this acidic precipitation, by acidification of lakes, has been responsible for extinction of acid-sensitive aquatic species and disruption of primary production and the nutritional food web within the affected ecosystems. Direct and indirect injury to crops and other vegetation by acidic precipitation has been postulated based on laboratory, greenhouse, and field experiments. However, some studies have revealed beneficial responses of crops exposed to acidic precipitation. In addition, acidic precipitation has the potential to produce deterioration of stone buildings, monuments, and a variety of other materials. There is also some concern about the impact of acidic precipitation on human health, although this impact is probably remote.

This section discusses the potential impacts of acidification on the environment. It must be recognized throughout the discussion that there are broad gaps in the data on which many of the assumed impacts are based, and there should be a clear distinction between what is known and what is unresolved. Most of the data available to date on impacts of acidic precipitation are derived from studies of the effects of increased acidity on aquatic organisms. The effects of lowering pH on fish, plant species, and other members of freshwater ecosystems are well documented; therefore, the manner and severity of disruption of the affected aquatic ecosystems produced by acidification may be postulated with some confidence.

Discussion of potential impacts of acidic precipitation on terrestrial ecology rests on more tenuous evidence. Most of these data were generated under laboratory or greenhouse conditions using simulations of exposure of terrestrial species to acidic precipitation. Therefore, any conclusions drawn from these data suffer by extrapolation from the laboratory to the field. This is not a refutation or condemnation of laboratory-generated data; however, the limitations of extrapolating these data, especially to as complex a system as a forest or other terrestrial ecosystem, must be recognized. Also, any generalizations or postulations on the impacts of acidic precipitation suffer from the fact that, to date, there has been no visible or detectable damage to terrestrial ecosystems outside the laboratory. This may be contrasted with the aquatic ecosystems where actual acidification-induced impacts have been observed and measured in situ.

Finally, it must be recognized that there has been no clear agreement among researchers regarding quantification of the magnitude of the potential adverse impacts of acidic precipitation, and whether the observed effects are a local or regional phenomenon caused by poor buffering capacity of the affected lakes or soils or whether the effects are more widespread.[11a] This question is difficult to resolve at present because research has centered on those areas where effects, especially aquatic effects, have been observed. Such research has therefore involved the most acid-sensitive regions, systems, and organisms. Implied impacts of acid precipitation on more highly buffered areas whose acid resistance is higher are somewhat speculative at this point.

EFFECT OF ACIDIC PRECIPITATION ON AQUATIC ECOSYSTEMS

Acidification of Lakes

The increasing acidity of freshwater lakes and streams appears to be a major environmental factor stressing aquatic ecosystems in Europe and North America.[11b]

The chemical composition of lakes is largely determined by the composition of influents from precipitation and watershed drainage. It has been postulated that the pH of some lakes may be affected by fertilizer runoff. Although this may well be true, the lakes in the Adirondacks where acidification has been reported typically are not only remote from any agricultural activity but are also at higher elevations. The complex pathways contributing to acidic inputs to lakes and streams are illustrated in Figure 4-1. Softwater lakes are usually produced by drainage of acidic igneous rocks, whereas hard waters contain large concentrations of alkaline earths derived from drainage of calcareous deposits.[12]

The acidity of freshwater lakes reflects both the acidity of precipitation and the capability of the watershed and the lake itself to neutralize incoming acid.[13] Schofield has remarked[14,15] that the situation is analogous to a large-scale chemical titration. The titration endpoint for each lake affected is determined by its hydrology and the capacity of the soils in the drainage basin to neutralize the incoming acid.[16] As noted, hardness of water is associated with alkalinity and, therefore, with the increased capacity of the water to neutralize or buffer the acidity entering a lake.[13] Chemical weathering and ion exchange are two mechanisms in watersheds that act to neutralize incoming acidity. The rate at which these processes proceed is dependent on the physical and chemical nature of the bedrock and soils.

The bicarbonate ion provides most of the buffering capacity of softwater lakes. The concentration of bicarbonate in soft waters is highly pH-dependent, and as the pH of precipitation and runoff decreases, the bicarbonate concentration also decreases. In acidified softwater lakes, sulfate, largely supplied by acidic precipitation, replaces bicarbonate as the major anion.[13,17,18] Bicarbonate in these lakes is essentially eliminated, and no effective

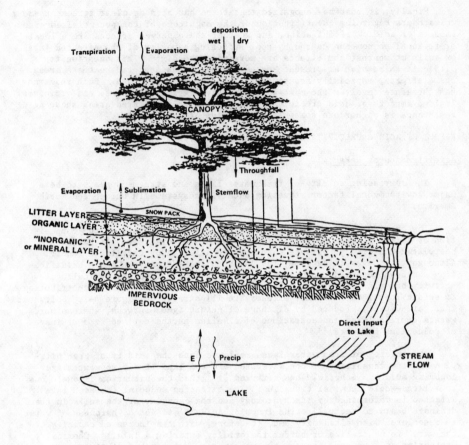

Figure 4-1. Definition sketch for the pathways of water tributary to a lake.

buffering capacity remains. These poorly buffered waters are subject to large fluctuations in acidity, especially in response to the large influxes of acidic pollutants observed following melting of ice and snow.[13-15,19,20] These sudden episodes of acidic input have been the cause of fish kills that occur during the late winter and early spring thaw; fish are often unable to adapt to the abrupt change in pH.[13,14,18,20-23] In areas where the watersheds or lake waters do not have any capability remaining to neutralize incoming acid, through loss of bicarbonate buffering, acidic precipitation causes the pH of the lake to drop permanently below 5.0. Fish have usually been observed to disappear from these waters.

Depending on various factors, therefore, lakes exhibit a range of sensitivity to acidification. Included are the acidity of both wet and dry atmospheric deposition, the hydrology of the lake, the soil system, and the resultant chemistry of the surface water. Among these factors, the most important in assessing the impact of acidic precipitation on lake acidification appears to be the soil system and associated canopy effects relative to the lake in question. Studies indicate that the capability of a lake and its drainage basin to neutralize the acidic inputs of precipitation is largely predicated by the composition of the bedrock of the watershed.[12,13,24] Lakes vulnerable to acidic precipitation have been shown to have watersheds whose geological composition make them resistant to chemical weathering.[13,24,25] Also, watersheds of acidic-sensitive lakes usually have poor soils with thin vegetation. The cation-ion exchange capacity of such soils is poor. Weathering and ion-exchange have been shown to be two crucial mechanisms by which acidity input to lakes may be neutralized.[14,24]

As shown, therefore, bedrock geology is generally a good measure of the susceptibility of an area to acidification caused by acidic precipitation. Using bedrock geology as an indicator, Galloway and Cowling[25] mapped those areas of North America that have the potential for being sensitive to acidic precipitation (see Figure 4-2).

The shaded areas in Figure 4-2 indicate bedrock composed of igneous or metamorphic rock, whereas unshaded areas are calcareous or sedimentary rock. Igneous or metamorphic bedrock weathers slowly, therefore, lakes in these areas would be assumed to have low alkalinity and low buffering capacity. Galloway and Cowling verified this assumption by compiling alkalinity data; lakes with low alkalinity were consistently found in regions having igneous and metamorphic bedrock.[25]

Although bedrock geology is a good indicator of susceptibility of an area to acidification, other factors exert an influence as well. 'As an example, there are areas in Maine with granite bedrock, which commonly has a low capacity for buffering, where lakes have not become acidified despite receiving precipitation with an average pH of about 4.3. It is believed that acidification has not occurred because drainage basins feeding these lakes contain lime-bearing till and marine clay. Small quantities of limestone in a drainage basin can apparently exert a strong influence on water quality in areas that would otherwise be assumed susceptible to acidic precipitation. Attempts to classify regions by sensitivity to acidic precipitation must take into account the potential buffering capacity of rock mixtures and the types of soils overlying the rock formations.

Figure 4-2. Regions in North America with lakes which may be sensitive to acid precipitation, using bedrock geology as an indicator.[25]

It must be emphasized that although the mechanisms of acidic input into freshwater lakes and streams have been recognized, the magnitude of the contribution of acidic precipitation to lake acidification is far from resolved. Many studies have emphasized the complex nature of the interactions between precipitation and resultant water quality. Some authors caution that water quality effects; i.e., acidification, usually attributed directly to the input of acidic precipitation, could possibly be the result of lithospheric or ecosystem changes not caused by acid deposition.[26] Some European investigators assign a secondary role to acidic precipitation in water quality changes. Rosenqvist concluded that the acidity of soil leachate was a factor determined more by patterns of agricultural land use than by acidic precipitation.[27,28] Others maintain that acidic precipitation is the causative factor.[28]

These differences remain to be resolved, and a complete assessment of the magnitude of the effects of acidic precipitation on freshwater quality is needed. This assessment requires detailed modeling as well as field and laboratory investigation. Among questions to be answered are as follows.

- What is the true importance of bedrock geology on the impact of acidic precipitation on water quality?

- What is the impact of possible mitigating chemical or other mechanisms in surrounding soil and rock mixtures and how may these factors be quantified?

- What is the impact of acidic precipitation on fish declines in sensitive areas?

This study would have to include consideration of all other factors potentially responsible, including increased recreational use, decrease in lake stocking, impact of agricultural fertilizer runoff, etc. Such a study would also require detailed chemical analyses and the chemical history of the water bodies surveyed. Finally, a quantification of the extent of the determined impacts of acidic precipitation has to be attempted. Is it a truly regional problem only affecting areas particularly sensitive, or is the impact more widespread, economically or otherwise? These areas all await systematic scientific investigation.

Effects on Fish

The death of fish in acidified freshwater lakes and streams has been more thoroughly studied, both in the laboratory and in the field, than any other aspect of lake and stream acidification. Various factors that affect the tolerance of fish to acidic waters have been identified, among which are species, strain, age, and size of the fish and physical factors including temperature, season, and hydrology.[1] Species of fish vary in their tolerance to low pH. Among the salmonids, rainbow trout are most sensitive, salmon are next, and brown and brook trout are least sensitive.[1] These data are based on experiments conducted with fish maintained at constant pH. Data are not available on species' response to transient pH changes.

Strains of the same species have demonstrated differing survival times either through variation in acclimative ability or genetics. Laboratory data reveal that larger fish are more tolerant than those which are younger.[1] The sensitivity of fish eggs has also been shown to vary with species.[1]

Laboratory studies have indicated that the higher the water temperature, the shorter the survival time. Survival of a species has also been found to vary with the season. Thus, acid episodes occurring at differing times of the year or at different water temperatures may differ in toxic effects.[1]

Chemical factors also influence fish survival in acidic waters. These factors include the ionic composition of the water, synergism or antagonism of toxic ions and the presence of toxic organics. Table 4-1 presents a summary of effects of pH changes on fish (adapted from Reference 29).

The decline of fish populations in acidified lakes and streams has been reported in Scandinavia,[21,30-35] Canada,[36] and more recently in the United States.[22] Although the disappearance of fish populations in Scandinavia was initially reported as long as 50 years ago, the rate of such disappearances has sharply increased during the past 15 years. Surveys conducted in southern Norway indicate large portions of the fish populations, especially trout and salmon, have been adversely affected by acidification.[21] Similar changes have been observed in Sweden.[32,35] Beamish and co-workers have documented the acidification of lakes and the loss of fish in Sudbury, Ontario.[17,36] Surveys conducted in New York State's Adirondack Mountains[37,38] and in Pennsylvania have indicated an increase in the number of lakes and streams with acid pH (<5) over time.

Upon reanalysis of original data, one study which attempted to correlate a decrease in pH with decreasing fish populations found that the data provide no indication of any significant correlation between changes in pH and changes in number of fish species.[104] This study goes on to refute the potential link between alleged trends in decreasing pH and decreasing numbers of fish species and acid precipitation. Furthermore, it concludes that there is some documentation that, to a limited degree, past and present use of pesticides, particularly DDT and methoxychlor, are having a detrimental effect upon fish stocks in the Adirondack region. The toxic effects of these compounds tend to vary with: temperature, age of fish, duration of exposure, persistence of pesticide, and the ability of the pesticide to bioaccumulate.

Although the effects attributed to the extensive past usage of DDT should continue to diminish with the passage of time, the present use of methoxychlor appears to remain a problem. The magnitude of pesticide effects upon fish stocks cannot be quantified without further research. However, it appears that pesticide effects as well as decreased stocking efforts and shifts in land use activity may be contributing to the decline in fish stocks in the Adirondack lakes.[104]

Field surveys in Norway,[30,34] Sweden,[32] Canada,[29] and the United States[14,37] have indicated that most fish species disappear from acidified

TABLE 4-1. SUMMARY OF EFFECTS OF pH CHANGES ON FISH[29]

pH	Effects
11.5 – 11.0	Lethal to all fish.
11.5 – 10.5	Lethal to salmonids; lethal to carp, tench, goldfish, pike if prolonged.
10.5 – 10.0	Roach, salmonids survive short periods, but lethal if prolonged.
10.0 – 9.5	Slowly lethal to salmonids.
9.5 – 9.0	Harmful to salmonids, perch if persistent.
9.0 – 6.5	Harmless to most fish.
6.5 – 6.0	Significant reductions in egg hatchability and growth in brook trout under continued exposure.
6.0 – 5.0	Rainbow trout do not occur. Small populations of relatively few fish species found. Fathead minnow spawning reduced. Molluscs rare. Declines in a salmonid fishery can be expected. High aluminum concentrations may be present in certain waters causing fish toxicity.
5.0 – 4.5	Harmful to salmonid eggs and fry; harmful to common carp.
4.5 – 4.0	Harmful to salmonids, tench, bream, roach, goldfish, common carp; resistance increases with age. Pike can breed, but perch, bream, and roach cannot.
4.0 – 3.5	Lethal to salmonids. Roach, tench, perch, pike survive.
3.5 – 3.0	Toxic to most fish; some plants and invertebrates survive.

lakes when the pH drops below 5. This is most probably caused, as indicated by laboratory data, by reproductive inhibition.[14,33,34,37,40-44] Experimentation has indicated that fish eggs and fry are sensitive to acidic water.[33] Both inhibition of gonad maturation[17,40] and mortality of eggs and larvae[30,42,43] can contribute to reproductive failure in fish populations inhabiting acidified waters.

Disappearance of fish from affected bodies of water usually may be the result of two patterns. A sudden, short-term shift in pH resulting in acid shock may cause fish mortality. Sudden drops in pH could cause fish kills at pH levels above those normally toxic to fish.[23,34,43] Such pH shocks often occur in early spring when snow melt releases acidic constituents accumulated during the winter.

A gradual decrease in pH with time, rather than sudden acid shock, is a second mechanism whereby acidification of water bodies could result in elimination of fish populations. Based on field observations and laboratory experimentation, as noted above, prolonged acidity interferes with fish reproduction and spawning so that, over time, there is a decrease in fish population density and a shift in the size and age of the population to older and larger fish. This pattern has been observed in Norway,[22,23] Sweden,[32] Canada,[17,36] and the United States.[22,37] It is important to note that even small increases (5 to 50 percent) in mortality of fish eggs and fry can significantly decrease fish populations and eventually bring about the extinction of the species in the affected water body.[37]

The physiological mechanisms responsible for mortality of fish in acidic waters may vary in response to levels of acidity and the presence of such components as heavy metals and CO_2. At the pH levels usually encountered in acidified waters (4 to 5), disruption of osmoregulatory functions is the most likely cause of fish death.[21,37] Laboratory studies have revealed impaired sodium uptake in fish at the pH concentration range present when lakes become acidified.[16,19]

Studies in the Adirondacks have indicated that mobilization of toxic metals, especially aluminum, is an additional factor that may contribute to mortality of fish at low pH values.[14,29] Soil leaching and mineral weathering by acidic precipitation may result in high concentrations of aluminum in surface and ground waters. Also, increased transport of aluminum into aquatic systems may affect phosphorous availability.[45] Aluminum has been observed to be toxic at pH levels as low as 4.0. Gill damage is reported to be a symptom of aluminum toxicity, but it is not clear whether gill damage is the mechanism causing mortality and whether this damage is specific for aluminum. Manganese is also believed to be mobilized by acidic precipitation, but its toxic effects at low pH, if any, are unknown.[16] Several studies in Sweden,[46] Canada,[47,48] and the United States,[47] have revealed high mercury concentrations in fish from acidified regions. Tomlinson[47] has reported that precipitation is the source of high mercury levels in the Bell River area of Canada. The results of at least one study suggest that acid stress and terrestrial impacts of mercury to several lakes have enhanced mercury uptake in fish.[105] Elevated mercury levels in fish or freshwater lakes would pose potential concern for both aquatic species and human health impacts. However, studies done to date are far from comprehensive, and reports of results are still controversial. Mercury levels reported by Tomlinson[47] are an order of magnitude greater than U.S. drinking water standards. Data on effects of organic acids in acidified lakes and streams are also unavailable.

Effects on Plant Life and the Food Chain

Elimination or reduction in the fish population is the most obvious biological impact associated with acidification of freshwater lakes and streams. Less obvious, but of great importance, however, are the effects of acidification on other aquatic organisms. Organisms at all trophic levels within the food chain may be affected. Species can be reduced in number and variety, and primary production and decomposition may be impaired with a resultant disruption of the entire ecosystem.

Within the aquatic ecosystem, energy moves along two pathways, the grazing food chain and the detrital food chain.[49,50] The green plants, the primary producers (phytoplankton, mosses, algae, etc.), are the food base of the grazing food chain where plants are eaten by animals and animals by other animals. Decomposers (bacteria, fungi, some protozoa, etc.) use dead plant and animal matter as food and release minerals and other compounds back into the environment. Thus, in the detrital food chain, the base is dead organic matter. Disruption of either of these pathways could create a chain reaction resulting in complete disruption within the ecosystem.

Changes in pH have caused changes in the composition and structure of the aquatic plant communities involved in primary production. Lowering of pH of lakes studied in Ontario resulted in changes in species composition and in the standing crop and production of the phytoplankton community. In these lakes, for example, the species of Chlorophyta (green algae) were reduced in number from 26 to 5, the Chrysophyta (golden brown) from 22 to 5, and the Cyanophyta (blue-green) from 22 species to 10.[51] In addition, the relative abundances of the algal flora also changed. Differences in nutrient levels (phosphorous and nitrogen) were not responsible for these changes in primary productivity; acidity appeared to be the limiting factor.

A study of phytoplankton populations in 115 lakes in Sweden supports the findings of the Ontario study.[21] The species composition of lakes with a pH less than five lacked diversity, and the size of the population was restricted.[32] Similar data were reported in a regional survey of 55 lakes in southern Norway.[23]

Acidification of lakes also results in reduction in productivity and species diversity of the macrophyte community. Studies of macrophyte communities in six lakes in Sweden have indicated that in five of the six lakes, Lobelia communities are being replaced by Sphagnum.[52,53] The abundance of Sphagnum mats covering the lake bottoms, which chokes out Lobelia, has been positively correlated with decreasing pH. The consequences of the Sphagnum occupying increasing areas of these lake bottoms are several. Basic ions (e.g., Ca^{+2}) that are necessary for biological production are bound to the moss tissue because of its strong ion exchange capacity and are, therefore, unavailable to other organisms. In addition, the benthic habitat deteriorates because the Sphagnum filamentous algae such as Mougestia proliferate. This has been verified under laboratory conditions at pH 4.0.[21]

Heavy growths of these filamentous algae and mosses have also been reported in Norwegian streams affected by acidification. These effects have been observed in artificial stream channels in which water and naturally seeded algae from an acidified brook (pH 4.3 to 5.5) were used. Lowering the pH to 4.0 increased algae growth when compared to controls.[52,54]

Reductions in the diversity of the plant communities in lakes and streams and subsequent disruption in primary production reduces the supply of food and, therefore, the energy flow within the affected ecosystem. Changes in these communities also reduces the supply of minerals and nutrients. These factors limit the number of organisms that can exist within the ecosystem.

Effects on Microorganisms and Decomposition

Acidification of lakes reduces microbiological activity and, therefore, affects the rates of decomposition and the accumulation of organic matter in aquatic ecosystems. Organic matter (detritus) in lakes plays a central role in the energetics of lake ecosystems.[12] The biochemical transformations of detrital organic matter by microbial metabolism are fundamental to nutrient cycling and energy flux within the system, and the trophic relationships within lake ecosystems are almost entirely dependent on detrital structure.[12]

Bacteria are central to the food relationships in a lake. Experiment has shown that the populations of decomposer organisms change from bacteria to fungi as pH is lowered.[21,52] Interference with nutrient cycling through disruption of the detrital trophic structure could, therefore, be a major result of changes in microdecomposer populations caused by acidification.[52] Accumulation of organic litter observed in acidified Swedish lakes, produced by extensive mats of fungal growth, seals off nutrients that would otherwise be available if normal decomposition occurred.[55]

Effects of Other Aquatic Organisms

Invertebrate communities are also affected by acidification of freshwater lakes and streams. Surveys conducted at sites in Scandinavia and North America have shown that acidified lakes and streams have fewer species of benthic invertebrates than do waters with higher pH.[23,56-58] Zooplankton analyzed from samples collected from 84 Swedish lakes showed that acidification had caused limitation of many species and led to simplification of zooplankton communities.[32] The distributions of crustacean zooplankton in acidified lakes in Ontario were shown to be strongly related to pH. As acidity increased, the complexities of zooplankton communities decreased.[58a] At a pH of about 5, an abrupt change from complex to simple zooplankton

communities occurred. In some lakes only a single species was found.
Reduced diversity in zooplankton communities affects the food supply and
thus causes changes in the community structure of organisms dependent on
the zooplankton as a food source.

Gastropods are also affected by acidic water conditions. In a survey of Norwegian lakes no snails were found when the pH was less than
5.2.[21,59] The amphipod *Gammarus lacustris*, an important element in the
diet of trout in Norwegian lakes where it occurs, is not found in lakes
with a pH below 6.0.[21,52] Experiments have revealed that adults of this
species cannot tolerate two days of exposure to a pH of 5.0.[21,52] The
short-term acidification which often occurs during spring snow melt could
eliminate these species from small lakes.

The tolerance of aquatic invertebrates to low pH varies during their
life cycle. Adult insects seem to be particularly sensitive at emergence.
In general, although there is broad variation, available data indicate a pH
value of 5.5 or higher is necessary for 50 percent viable emergence.[60]

Amphibians may be the species most directly affected by acidic precipitation. The reproductive habits of amphibians make them especially susceptible to the pH changes of the ponds where they lay their eggs. Frogs,
toads, and most salamanders in the United States lay eggs in ponds.[61] In
addition, many species breed in shallow, temporary pools that are strongly
affected by the pH of the precipitation that fills them.[61]

Frog embryos have been observed to develop abnormally at pH 3.7 to 4.6,
and pH <4.0 is usually lethal.[62] A pH lower than 6.0 can inhibit the development and increase egg mortality of spotted salamanders.[63] Based on
available data, it appears that reproduction in amphibians, as in fish, is
primarily affected first when pH is lowered.[64]

Frogs and salamanders are important predators of invertebrates (mosquitoes, etc.) and they themselves are important prey for higher trophic levels
in the ecosystem. Increasing acidity in freshwater habitats results in
shifts in species, populations, and communities. Any species that depends
on aquatic organisms (plant or animal) for a portion of their food will be
affected. A summary of the changes most likely to occur in aquatic biota
with decreasing pH is presented in Table 4-2. Table 4-3 lists effects of
decreasing pH on aquatic organisms.

EFFECT OF ACIDIC PRECIPITATION ON TERRESTRIAL ECOSYSTEMS

Assessing the impacts of acid precipitation on terrestrial ecosystems
is extremely difficult. In aquatic systems, it has been possible to measure
pH changes and correlate the observed effects on aquatic species with the pH
shift. To date, it appears that no component of terrestrial ecosystems is
as sensitive to the impacts, if any, of acidic precipitation as is a poorly
buffered aquatic system. In fact, at present, it has not been possible to
observe or measure changes in natural terrestrial ecosystems that could be
directly attributed to acidic precipitation; however, such changes have been

TABLE 4-2. CHANGES IN AQUATIC BIOTA MOST LIKELY
TO OCCUR WITH INCREASING ACIDITY[65]

1. Bacterial decomposition is reduced and fungi dominate saprotrophic communities. Organic debris accumulates rapidly.

2. The ciliate fauna is greatly inhibited.

3. Nutrient salts are taken up by plants tolerant of low pH (mosses, filamentous algae) and by fungi. Thick mats of these materials may develop, inhibiting sediment-to-water nutrient exchange and choking out other aquatic plants.

4. Phytoplankton species diversity, biomass, and production are reduced.

5. Zooplankton and benthic invertebrate species diversity and biomass are reduced. Remaining benthic fauna consists of tubificids and *Chironomus* (midge) larvae in the sediments. Some tolerant species of stone flies and mayflies persist as does the alderfly. Air-breathing bugs (water-boatman, backswimmer, water strider) may become abundant.

6. Fish populations are reduced or eliminated.

7. Changes in populations and communities occur at virtually all trophic levels.

TABLE 4-3. SUMMARY OF EFFECTS OF AQUATIC ORGANISMS WITH DECREASING pH[65]

8.0-6.0	• Long-term changes of less than 0.5 pH units in the range 8.0 to 6.0 are very likely to alter the biotic composition of freshwaters to some degree. The significance of these slight changes is, however, not great.
	• A decrease of 0.5 to 1.0 pH units in the range 8.0 to 6.0 may cause detectable alterations in community composition. Productivity of competing organisms will vary. Some species will be eliminated.
6.0-5.5	• Decreasing pH from 6.0 to 5.5 will cause a reduction in species numbers and, among remaining species, significant alterations in ability to withstand stress. Reproduction of some salamander species is impaired.
5.5-5.0	• Below pH 5.5, numbers and diversity of species will be reduced. Many species will be eliminated. Crustacean zooplankton, phytoplankton, molluscs, amphipods, most mayfly species, and many stone fly species will begin to drop out. In contrast, several pH-tolerant invertebrates will become abundant, especially the air-breathing forms (e.g., Gyrinidae, Notonectidae, Corixidae), those with tough cuticles that prevent ion losses (e.g., _Sialis lutaria_), and some forms that live within the sediments (Oligochaeta, Chiromomidae, and Tubificidae). Overall, invertebrate biomass will be greatly reduced.
5.0-4.5	• Below pH 5.0, decomposition of organic detritus will be severely impaired. Autochthonous and allochthonous debris will accumulate rapidly. Most fish species will be eliminated.
4.5 and below	• Below pH 4.5 all of the above changes will be greatly exacerbated. Lower limit for many algal species.

observed under controlled laboratory and field conditions. Therefore, it may be postulated that such effects could occur. Although this postulation is somewhat tenuous, assuming as it does extrapolation from the laboratory to the natural situation, it is probably prudent to view these potential effects as possible, even as a worst case. Such worst-case assumptions are frequently made to assess potential impacts on health or the environment, especially where the capability of the observer to measure the effect is limited by the complexity of the natural situation or because the impact is only fully realized after a long time (e.g., decades).

One study indicates that nitrates are usually in short supply during the growing season, and both terrestrial and aquatic ecosystems take up nitrates during this time. This conclusion would seem to indicate that nitrates pose little or no problem to terrestrial ecosystems.

Effects on Vegetation

Chemical species in the atmosphere reach plant surfaces through wet and dry deposition. Although sulfates, nitrates, and other water-soluble species may be assimilated through plant leaves, it has generally been assumed that the free hydrogen ion concentration in acidic precipitation is the component most likely to cause direct, harmful effects on vegetation. Experimental studies have supported this assumption, but, as noted above, there have been no reports of foliar symptoms on field-grown vegetation in the continental United States that could be attributed to exposure to ambient acidic precipitation.[66] The most frequently reported response of vegetation to experimental exposure of simulated acidic rain is the formation of lesions or areas of dead tissue on leaf surfaces.[66,67] A large percentage of the leaf area may exhibit such lesions after repeated exposures to simulated acid rain at pH levels of 3.0, 3.7, 2.5 and 2.3.[68,69] Pinto bean leaves exhibited pronounced leaf injury.[68,70,71] Most leaf injury caused by exposure to simulated acid rain has been observed to occur on expanding or recently expanding leaves.

A recent study, which is now partially completed, measured the effects of acid precipitation on 32 major crops that represent a total annual income in the United States of $50 billion. Crops were grown under controlled environmental conditions and exposure to simulated (sulfuric) acid rain of pH: 3.0, 3.5, and 4.0, in addition to a control rain of pH 5.7. Injury to foliage and effects on yield of edible portions were then determined.[72]

Preliminary results indicated that some crops suffered severe damage, but others sustained little apparent injury. The leafy portions of mustard greens and the edible portions of broccoli exposed to simulated rain of pH 3.0 were reduced in weight an average of 30 percent and 25 percent, respectively, as compared to controls. The edible portions of radishes exposed to pH 3.0 weighed less than half that of radishes receiving control rain. Spinach growth was reduced only 15 percent, but the leaves were so badly pitted that the spinach would be unmarketable.[72]

It should be mentioned that exposure of crops to pH <3.5 represents a situation which probably would not exist in crop-growing areas.

Under less acidic conditions (pH 3.5-4.0) yields of mustard greens, radishes, and bluegrass were 14-28 percent less than corresponding controls. Cauliflower, cabbage, and green peas, however, were neglibly affected under any experimental conditions.[72]

In leaves injured by simulated acidic rain exposure, collapse and distortion of cells on the upper leaf surface is frequently followed by further injury until all leaf surfaces are affected.[73] Although foliar injury is commonly observed, reductions in growth and crop yield have not been unequivocally associated with leaf injury. In addition, foliar response to exposure to acidic precipitation has been shown to be dependent on numerous factors. Duration and frequency of exposure, acid content and size of rain drops, and the intensity of rainfall all influence the development and extent of foliar symptoms.[74] Environmental conditions during and after precipitation may affect the response of plants. Variation in such conditions could alter physiological processes, the amount of liquid remaining on leaf surface, the rate of evaporation after the rainfall has stopped, etc.[74] Recognition and appreciation of these variables accentuate the difficulties in extrapolating laboratory results to the natural condition. Table 4-4 summarizes available results on direct injury to vegetation by exposure to simulated acidic precipitation.

Acidic precipitation can also cause indirect effects on plants and vegetation, some of them beneficial. Wood and Bormann reported an increase in needle length and the weight of seedlings of Eastern white pine with increasing acidity of simulated precipitation.[67] Investigators at the Argonne National Laboratory have reported no harmful effects on soybean productivity following exposure to simulated acidic rain. In fact, they observed a positive effect on productivity as reflected by seed growth.[81] As noted above, an Environmental Protection Agency study showed negligible effects of acid precipitation on some exposed crops.[72]

Shriner[82] reported on the effects of acidic precipitation on host-parasite interactions. Simulated acid rain with a pH of 3.4 inhibited the development of bean rust. Response of other diseases such as halo blight of bean seedlings was observed to vary depending on the time in the disease cycle during which simulated acidic rain was applied.

The effects of acidic precipitation on forest tree growth have been investigated. Studies have been conducted using simulated acid rain and by comparing development of the rings in the past with present ring formation.[83] Jonsson and Sundberg[84] used this tree ring method to analyze forest growth in southern Sweden from 1896 to 1965. They report a 2 to 7 percent decrease in growth between 1950 and 1965, which they attribute to increases in acidification. Professor Folke Andersson, Swedish University of Agriculture, has reported that these ring studies have recently been updated, and the trends that were previously reported are no longer observed. Studies in Norway and North America could not corroborate their findings,

TABLE 4-4. SUMMARY OF DIRECT FOLIAR INJURY FOLLOWING EXPOSURE TO SIMULATED ACIDIC PRECIPITATION

Effect	Receptor	pH value	Reference
Foliar lesions	Eastern white pine	2.3	69
Foliar aberrations decrease in growth	Bean	2.5	75
Foliar lesions, plasmolysis of cells, reduction in dry weight	Bean	2.5	69
Foliar necrotic spots	Scots pine birch (Bethula pubescens Ehrb)	2.5	76
Necrotic lesions and chlorotic areas in leaves	Soybean	3.0	77
Foliar lesions, decrease in growth	Yellow birch (Betula alleghaniensis Britt)	3.1	78
Foliar lesions	Bean, sunflower	3.1	68
Bifacial necrosis	Oak (Quercus phellos)	3.2	79
Foliar lesions	Hybrid poplar	3.4	76
Foliar lesions	Sunflower	3.4	66
Reduction in dry weight	Bean	4.0	80
Reduction in dry weight	Mustard greens	3.0	72
Reduction in dry weight	Broccoli	3.0	72
Reduction in dry weight	Radishes	3.0	72
Foliar lesions, reduction in dry weight	Spinach	3.0	72
Decrease in growth	Mustard greens, radishes	3.5-4.0	72

however.[76,85,86] In fact, simulated acidic precipitation was observed to increase the growth of pine saplings in experiments conducted in Norway.[83] Saplings in test plots watered with acidic rain of pH 3.0, 2.5, and 2.0 grew more rapidly than did control trees.

Among other postulated effects of acidic precipitation on terrestrial systems are increased leaching of chemical elements from exposed plant surfaces and/or forest soils. Leaching of organic and inorganic materials from vegetation to the soil is part of the natural functioning of terrestrial ecosystems and forms an integral phase in nutrient cycling. Plant leachates affect soil texture, aeration, permeability, and ion-exchange capacity. These leachates influence the number and behavior of soil microorganisms and thereby soil fertility and the immunity of plants to pests and disease.[87]

Research to date on the effects of acidic precipitation on the leaching of chemical constituents from vegetation has resulted in equivocal and often contradictory results. It has been demonstrated that acidic precipitation can increase the leaching of various cations and organic carbon from the tree canopy.[76,88] Foliar losses of nutrient cations from bean plants and maple seedlings were found to increase as acidity of the artificial mist to which they were exposed increased. However, in experiments using Norway spruce, researchers found no evidence of change in the foliar cation content although increased leaching was observed.[83] It has been stated that increased leaching of nutrients from foliage can actually accelerate their uptake by plants.[87] The impact of the increased leaching of chemical substances from vegetation by acidic precipitation is still unresolved.

Increased acidic precipitation may cause a decrease in the fertility of forest soils. Laboratory studies have shown that leaching of the important nutrients potassium, magnesium, and calcium is accelerated by increased acidity.[89] However, other workers contend that soil fertility may be increased as a result of deposition of nitrate and sulfates (typical components of chemical fertilizers) in the acidic rain. Finally, another major uncertainty in estimating effects of increased acidic precipitation on forest fertility and productivity is the as-yet unquantified capability of forest soils to buffer against leaching by hydrogen ions.

Effects on Soil

Another area that suffers from limited investigation and equivocal results is the effects or consequences of increased acidity on soil and the subterranean ecosystem. Effects have been postulated, but the picture is far from clear. It is especially difficult to factor out the impacts of acid precipitation on soil, if any, as compared to natural or anthropogenic mechanisms resulting in soil acidification, such as agricultural fertilization. Some authors contend that acid precipitation inputs to date are low compared to the possible influences of agricultural fertilization or liming practices.[90-92]

Soils may be exposed to (1) the influence of modified cation exchange, which result from the penetration of acidic precipitation, resulting in the losses of such species as Ca^{+2}, Mg^{+2}, K^+, and Na^+ and (2) the collection of metals that could result in soil contamination and potential phytotoxicity. As water containing H^+ moves through soil, some of the hydrogen cations replaced adsorbed exchangeable cations such as Ca^{+2}, Mg^{+2}, K^+, and Na^+, which may eventually be leached into ground water. All soils are not equally susceptible to acidification. The buffering capacity of soil depends on mineral content, texture, structure, pH, base saturation, salt content, and soil permeability. Buffering capacity is greatest in soils derived from sedimentary rocks, especially those containing carbonates, and least in soils derived from crystalline rocks such as granites and quartzites.[93] Soil buffering capacity varies widely in different regions of the country. The sensitivity of different soils based on pH, texture, and calcite content is summarized in Table 4-5.[94]

TABLE 4-5. THE SENSITIVITY TO ACID PRECIPITATION BASED ON BUFFER CAPACITY AGAINST pH CHANGE, RETENTION OF H^+, AND ADVERSE EFFECTS ON SOILS[94]

	Calcareous soils	Noncalcareous		Cultivated soils pH >5	Acid soils pH >5
		Clays pH >6	Sandy soils pH >6		
Buffering	Very high	High	Low	High	Moderate
H^+ retention	Maximal	Great	Great	Great	Slight
Adverse effects	None	Moderate	Considerable	None - slight	Slight

Few comprehensive studies have been conducted to assess the potential impacts of acidic precipitation on cation exchange and leaching on nutrients. Those which have been conducted tend to concur that increases in acidification of precipitation lead to loss of cation exchange capacity and increased rates of mineral loss (especially Ca^{+2}).[89,94,95]

Although the potential effects of acidic precipitation on soil could be long lasting, researchers note that many counteracting forces could mitigate the overall final effects, including the release of new cations to exchange sites by weathering or through nutrient recycling by vegetation.[96] Leaching of soil nutrients is efficiently inhibited by vegetation growing in and on the soil. Plant roots frequently absorb nutrients in amounts larger than the plants require. Large amounts of these nutrients will be deposited later on the soil surface as litter or as leachate from the vegetation canopy.[83]

Lowered soil pH can also affect the mobility and availability to plants of toxic metallic species. In general, the availability of toxic metals increases as pH decreases. Very low soil pH has been associated with the mobility of toxic aluminum compounds in soil. Ulrich[97] has reported that aluminum released by acidified soils could, in time, reach levels that would be phytotoxic. The effects of acidification on release of toxic metals is an area that is currently receiving increased research effort. Biological processes in the soil necessary for plant growth could also potentially be affected by soil acidification, including nitrogen fixation.[83] However, there are few data available in this area.

It is difficult to generalize about potential effects of soil acidification on microorganisms, as well. Many microbial processes that are essential for plant growth are suppressed as acidity increases; however, the inhibition observed in one soil at a given pH may not be seen at the same pH in another soil.[98] Also, studies performed to date have typically tested soils maintained for short periods of time at very low pH.

In summary, at present, there is little evidence of visible or even detectable damage to terrestrial ecosystems caused by acidic precipitation. Terrestrial ecology is a complicated biological system, and deposition of acidic precipitation exerts a complex influence on the functioning of that ecology. The evaluation of potential impacts is complicated by the apparent trade-offs between benefits from nutrient enhancement and the possibility of inhibition of plant growth or other detrimental effects. Although results to date have been generated in the laboratory under controlled conditions and no evidence of crop damage in the field has been observed, it is possible that acidic precipitation is producing or will produce some responses within the ecosystem even though it is not possible at the present time to observe, record, or evaluate them. The determination of what these changes are, the quantification of these changes, and the determination of whether the changes are harmful or beneficial can only be accomplished with certainty through systematic scientific investigation over a long period of time.

EFFECT OF ACIDIC PRECIPITATION ON MATERIALS

Acidic precipitation can damage materials, structures, and man-made artifacts. It has the potential to accelerate corrosion of metals and erosion of stone. However, because a dominant factor in the formation of acidic precipitation is sulfur species (especially SO_2), it is difficult to distinguish effects of acidic precipitation from damage induced by sulfur pollution in general. Laboratory and field studies that attempted to assess the effects of single pollutants and combinations of pollutants on materials at various concentrations have been conducted. Although physical damage functions for some pollutant species have been developed, most are based on exposure data collected at relatively high pollutant levels. Data from exposures to pollutants at ambient levels are generally unavailable. Generalizations based on laboratory data must, therefore, be regarded as somewhat tenuous. Field observations and measurements may be of greater value in determining the impact of acid pollutants on materials and structures, but isolation of these impacts from the impacts associated with other constituents of the ambient atmosphere is extremely difficult.

The influence of acidic precipitation on corrosion of metals has been investigated. Precipitation, as rain, can have varying influence on corrosion. Rain may accelerate corrosion by forming a layer of moisture on the metallic surface and by adding hydrogen and sulfate ions. However, rain may also wash away sulfates deposited during dry deposition and can, therefore, retard corrosion. Kucera[99] has investigated this problem and has concluded, as indicated, that mode of deposition complicates analysis of the impact of acid precipitation. It has been observed that in an area where dry deposition of hydrogen and sulfate ions exceeded that by wet deposition, flat steel plates corroded more rapidly on their undersides than on their upper, rain-exposed surfaces. This illustrates the washing effect of rainfall. However, in areas where wet and dry deposition were equivalent, the upper sides of the plates corroded more rapidly, indicating the predominance of rainfall's corrosive effects. Other variables were also observed to influence the impact of acid precipitation on corrosion, including amount and frequency of precipitation, pH, relative humidity, and temperature.[36]

High acidity in rainfall is believed to promote corrosion because the hydrogen ions present act as a sink for the electrons liberated during the corrosion process.[100] The metals most likely to be corroded by precipitation with low pH are those whose corrosion resistance depends on a layer of carbonates, sulfates, or oxides, such as zinc or copper. A pH of 4 or less in rainfall could accelerate the dissolution of these protective layers.[99]

Besides metals, limestone, sandstone, concrete, cement-lime, and lime plaster are reported to be adversely affected by acidic precipitation.[101] Sulfur compounds in the ambient atmosphere react with the carbonates in limestones and dolomites, calcareous sandstone, and mortars to form calcium sulfate. This reaction results in blistering, scaling, and loss of surface cohesion.[102]

Acidic precipitation may leach chemical constituents from stonework just as acidic water leaches ions from soils and bedrock. However, at the present time, it is not possible to attribute observed effects of atmospheric sulfur compounds in general, or acidic precipitation in particular, to specific chemical compounds. The precise chemical mechanisms involved in such deteriorations are, likewise, unresolved. However, the effects are evident on buildings, monuments, and statuary.[101,102]

EFFECTS OF ACIDIC PRECIPITATION ON HUMAN HEALTH

As previously observed, mobility of metallic compounds in soil is increased at low pH values. Given this fact, there exists a potential indirect impact on human health through contamination of edible fish and drinking water supplies by these metallic species. Although this impact is highly speculative, and data are scarce or nonexistent, the potential does exist and thus should not be ignored.

Changes in the concentrations of heavy metals have been observed in waters and aquatic species in acidified regions.[46-48] Increased concentrations of aluminum, manganese, zinc, copper, cadmium, and nickel have been reported.[13,17,103] Elevated mercury concentrations have also been observed. Beamish[12] has suggested that increased concentrations of nickel and copper in Canadian lakes studied could be the result of increased acidic precipitation, although copper concentrations in acidified lakes were not significantly different from those in nonacidified lakes. Likewise, although zinc concentrations were higher in acidified lakes as compared to nonacidified, the levels were well within the range reported from analysis of samples obtained from 1500 water bodies within the United States.[17]

Aluminum appears to be the primary element mobilized by acidic precipitation in regions characterized by soils with poor acid-buffering capacity. Cronan[45] found that acid deposition has apparently triggered the release of aluminum from soil horizons that would normally immobilize this element. Such releases of aluminum could have detrimental effects on fish populations in acidified lakes. In addition, there is a possibility that ingestion of fish contaminated by aluminum or other metals could represent a health hazard. However, comprehensive study and analysis of toxic metals in commercial or recreational fish catches has yet to be conducted. Also, reported concentration levels of these metallic species in waters analyzed have been orders of magnitude below public health drinking water standards. The bioaccumulative potential of fish for these metallic species is unknown.

Another possible human health impact is the potential that, as drinking water supplies acidify through increased input from acid precipitation, levels of metal concentrations in these waters will approach limits. This increase in the concentrations of metallic species could be caused by increased watershed weathering or by leaching of metals from household plumbing. However, only preliminary and inconclusive investigations of these areas have been made thus far.

SUMMARY

The data presented and discussed in this section suggest that acidic precipitation, by acidification of lakes, has been responsible for elimination of acid-sensitive aquatic species and has disrupted primary production and the nutritional food web within the affected ecosystems. These conclusions rely on an extensive data base relating the death of fish and other aquatic organisms to increased acidification of freshwater lakes and streams.

Discussion of potential impacts of acidic precipitation on terrestrial ecology rests on more tenuous evidence. To date, there has been no visible or detectable damage to terrestrial ecosystems outside the laboratory. In fact, some studies have indicated indirect beneficial responses due to exposure of vegetation to acidic precipitation.

It is difficult to factor out the effects of acidic precipitation on structures and materials from those of sulfur air pollutants in general, although these effects are evident on masonry and elsewhere.

Although the possibility exists that acid precipitation may have indirect, adverse impacts on human health, no comprehensive data are available, and any proposed relationships are currently considered speculative.

Finally, it must be recognized that there remain broad gaps in the data on which many of the assumed impacts of acidification are based. Also, there has been no clear agreement among researchers regarding the magnitude of the potential adverse impacts of acidic precipitation, nor whether observed effects are a local or regional phenomenon caused by poor buffering capacity of the affected lakes or soils or whether the effects are more widespread.

REFERENCES

1. Wood, T. Acid Precipitation. In: Sulfur in the Environment, Missouri Botanical Garden in cooperation with Union Electric Company. St. Louis, Missouri, 1975. pp. 39-50.

2. Likens, G. E., R. F. Wright, J. N. Galloway, and T. J. Butler. Acid Rain. Scientific American, 241: 43-51, 1979.

3. Glass, N. R., G. E. Glass, and P. I. Rennie. Effects of Acid Precipitation. Environ Sci. and Tech., 13: 1350-1355, 1979.

4. Likens, G. E. Acid Precipitation. Chem. Eng. News, 54:(48):29-44, 1976.

5. Likens, G. E., F. H. Bormann, and N. M. Johnson. Acid Rain. Environment, 14:33-40, 1972.

6. Braekke, F. H., ed. Impact of Acid Precipitation on Forest and Freshwater Ecosystems in Norway. SNSF Project, Research Report FR 6/76, Oslo, Norway, 1976.

7. Cogbill, C. V., and G. E. Likens. Acid Precipitation in the Northeastern United States. Water Resour. Res., 10:1133-1137.

8. Cogbill, C. V. The History and Character of Acid Precipitation in Eastern North America. In: Proceedings of the First International Conference on Acid Precipitation and the Forest Ecosystem, May 12-15, 1975, Columbus, Ohio, L. S. Dochinger and T. A. Seliga, eds., U.S. Forest Service General Technical Report NE-23, U.S. Department of Agriculture, Forest Service, Northeastern Forest Experiment Station, Upper Darby, Pennsylvania, 1976. pp. 363-379.

9. Likens, G. E. The Chemistry of Precipitation in the Central Finger Lakes Region. Technical Report No. 50, Water Resources and Marine Science Center, Cornell University, Ithaca, New York, 1972.

10. National Research Council. Air Quality and Stationary Source Emission Control. Commission on Natural Resources, National Academy of Sciences, for the Committee on Public Works, U.S. Senate, 94th Congress, 1st Session, Committee Serial No. 94-4, U.S. Government Printing Office, Washington, D.C., 1975.

11. Matheson, D. H., and F. C. Elder, eds. Proceedings of Symposium on Atmospheric Contribution to Chemistry of Lake Waters. J. Great Lakes Res., 2 (Suppl. 1):1-225, 1976.

11a. Hidy, G. M., D. A. Hansen, and R. C. Henry. Comments on External Review Draft No. 1 of the Air Quality Criteria for Particulate Matter and Sulfur Oxides. Prepared by Environmental Research and Technology, Westlake Village, Calif. for Prather, Seeger, Doolittle, and Farmer, Washington, D. C., July 1980.

11b. Hendry, G. R. (ed.). Luminological Aspects of Acid Precipitation. Proceedings of the International Workshop held at the Sagamore Lake Conference Center, September 25-28, 1978. Co-sponsored by Corvallis Environmental Research Laboratory (U.S. Environmental Protection Agency) and Brookhaven National Laboratory. Report No. BNL 51074.

12. Wetzel, R. G. Limnology. W. B. Saunders Co., Philadelphia, Pennsylvania, 1975. 743 pp., p. 287-521.

13. Wright, R. F., and E. T. Gjessing. Changes in the Chemical Composition of Lakes. Ambio., 5:219-223, 1976.

14. Schofield, C. L. The Acid Precipitation Phenomenon and its Impact in the Adirondack Mountains of New York State. In: Scientific Paper from the Public Meeting on Acid Precipitation, May 4-5, 1978, Lake Placid, New York. Science and Technology Staff, New York State Assembly, Albany, New York, March 1979. pp. 86-91.

15. Schofield, C. L. Acid Precipitation's Destructive Effects on Fish in the Adirondacks. New York Food Life Sci. Q., 10(3):12-15, 1977.

16. Pottes, W. T. W., and D. J. A. Brown. Study Group 5 Discussions. D.5.1-D.5.5. In: Ecological Effects of Acid Precipitation, M. J. Wood, ed. Report of workshop held at Cally Hotel, Gatehouse-of-Fleet, Galloway, United Kingdom, 4-7 Sept. 1978. EPRI SOA77-403, Report No. EA-79-6-LD, December 1979.

17. Beamish, R. J. Acidification of Lakes in Canada by Acid Precipitation and the Resulting Effects on Fish. Water, Air and Soil Poll., 6:501-514, 1976.

18. Schofield, C. Effects of Acid Rain on Lakes. Presented at the American Society of Civil Eng. Convention, Boston, April 2, 1979. 13 pp.

19. Leivestard, H., and I. P. Muniz. Fish Kill at Low pH in a Norwegian River. Nature, 259:391-392, 1976.

20. Hagen, A., and A. Langeland. Polluted Snow in Southern Norway and the Effect of the Meltwater on Freshwater and Aquatic Organisms. Environ. Pollut., 5:45-57, 1973.

21. Leivestad, H., G. Hendrey, I. P. Muniz, and E. Snekvik: Effects of Acid Precipitation on Freshwater Organisms. In: Impact of Acid Precipitation on Forest and Freshwater Ecosystems in Norway, F. H. Braekke, ed. Research Report FR 6/76, 1432 Aas - NLH, Norway: SNSF Project Secretariat, 1976. pp. 87-111.

22. Schofield, C. L. Lake Acidification in the Adirondack Mountains of New York: Causes and Consequences. In: Proceedings of the First International Symposium on Acid Precipitation and the Forest Ecosystem, May 12-15, 1975, Columbus, Ohio, L. S. Dochinger and T. A. Seliga, eds. U.S. Forest Service General Technical Report NE-23, U.S. Department of

Agriculture, Forest Service, Northeastern Forest Experiment Station, Upper Darby, Pennsylvania, 1976. p. 477. (Abstract).

23. Hendrey, G. R., and R. F. Wright. Acid Precipitation in Norway: Effects on Aquatic Fauna. J. Great Lakes Res., 2 (Suppl. 1):192-207, 1976.

24. Wright, R. F., and A. Henksen. Chemistry of Small Norwegian Lakes with Special Reference to Acid Precipitation. Limnol. Oceanog., 23:487-498, 1978.

25. Galloway, J. N., and E. B. Cowling. The Effects of Precipitation on Aquatic and Terrestrial Ecosystems: A Proposed Precipitation Network. JAPCA, 28:229-235, 1978.

26. Davis, R. B., M. O. Smith, J. H. Baily, and S. A. Norton. Acidification of Maine (U.S.A.) Lakes by Acidic Precipitation. Verh. Internat. Verein Limnol., 20:532-537, 1978.

27. Rosenqvist, I. T. A Contribution Towards Analysis of Buffer Properties of Geological Materials Against Strong Acids in Precipitation Water. Norwegian General Sci. Res. Council, Council for Res. and Natural Sci., Oslo, Norway, 1976. p. 99.

28. Holden, A. V., and J. F. Spencer. Study Group 4 Discussions. D4.1-D4.5. In: Ecological Effects of Acid Precipitation. M. J. Wood, ed. Report of workshop held at Calley Hotc, Catchouse-of-Fleet, Galloway, United Kingdom, 4-7 September 1978, EPRI SOA77-403 Report No. EA-79-6-LD, December 1979.

29. European Inland Fisheries Advisory Committee (EIFAC). Water Quality Criteria for European Freshwater Fish. Water Res., 3:596-611, 1969.

30. Wright, R. F., and E. Snekvik. Acid Precipitation: Chemistry and Fish Populations in 700 Lakes in Southernmost Norway. Verh. Int. Ver. Theor. Angew. Limnol., 20:765-775, 1978.

31. National Research Council. Sulfur Oxides. In: Effects of Sulfur Oxides on Aquatic Ecosystems, National Academy of Sciences, Washington, D.C., 1978.

32. Almer, B., W. Dickson, C. Ekstrom, E. Hornstrom, and U. Miller. Effects of Acidification on Swedish Lakes. Ambio., 3:30-36, 1974.

33. Jensen, K. W., and E. Snekvik. Low pH Levels Wipe Out Salmon and Trout Populations in Southernmost Norway. Ambio., 1:223-225, 1972.

34. Wright, R. F., T. Dale, E. T. Gjessing, G. R. Hendry, A. Henriksen, M. Johannessen, and I. P. Muniz. Impact of Acid Precipitation on Freshwater Ecosystems in Norway. Water, Air and Soil Poll., 6:483-499, 1976.

35. Dickson, W. The Acidification of Swedish Lakes. Fishery Board of Sweden, Institute of Freshwater Research, Drottningholm, Sweden. Report No. 54, Lund, Sweden, Carl Bloms Boktryckeru A.OB, 1975. pp. 8-20.

36. Beamish, R. J., and H. H. Harvey. Acidification of the LaCloche Mountain Lakes, Ontario, and Resulting Fish Mortalities. Jour. Fisheries Res. Board Canada, 29:1131-1143, 1972.

37. Schofield, C. L. Acid Precipitation: Effects on Fish. Ambio., 5:228-230, 1976.

38. Pfeiffer, M. List and Summary of Acidified Adirondack Waters Based on Data Available as of April 1979. New York State Department of Environmental Conservation, Albany, New York, May 1979.

39. Arnold, D. E., R. W. Light, and V. J. Dymond. Probable Effects of Acid Precipitation on Pennsylvania Waters. EPA-600/3-80-D12. U.S. Environmental Protection Agency, Corvallis, Oregon, 1980. 19 p.

40. Beamish, R. J., W. L. Lockhart, J. C. Van Loon, and H. H. Harvey. Long-Term Acidification of a Lake and Resulting Effects on Fishes. Ambio., 4:98-102, 1975.

41. Menendex, R. Chronic Effects of Reduced pH on Brook Trout (Salvelinus fontinalis). J. Fish. Res. Board Can., 23:118-123, 1976.

42. Trojnar, J. R. Egg and Larval Survival of White Suckers (Catostomus commersoni) at Low pH. J. Fish Res. Board Can., 34:262-266, 1977.

43. Trojnar, J. R. Egg Hatchability and Tolerance of Brook Trout (Salvelinus fontinalis) Fry at Low pH. J. Fish Res. Board Can., 34:574-579, 1977.

44. Schofield, C. L. Lake Acidification in the Adirondack Mountains of New York: Causes and Consequences. In: First International Symposium on Acid Precipitation and the Forest Ecosystem, USDA Forest Service, Northeastern Forest Experiment Station and Ohio State University Atmospheric Science Program, Columbus, Ohio, 1975. p. 25.

45. Cronan, C. S., and C. L. Schofield. Aluminum Leaching Response to Acid Precipitation: Effects on High-Elevation Watersheds in the Northeast. Science, 204:304-305, 1979.

46. Landner, L., and P. O. Larsson. Biological Effects of Mercury Fall-Out into Lakes from the Atmosphere. IUL Report B115. Institute for Water and Air Research, Stockholm, Sweden, 1972. 18 pp. (in Swedish, translated by H. Altosaar, Domtar Research Centre, December 25, 1975).

47. Tomlinson, G. H. Acidic Precipitation and Mercury in Canadian Lakes and Fish. In: Public Meeting on Acid Precipitation, May 4-5, 1978, Lake Placid, New York. Sponsored by the Committee on Environmental Conservation, New York State Assembly. Pub. by Science and Technology Staff, New York State Assembly. p. 104-118. March 1979.

48. Prouzes, R. J. P., R. A. N. McLean, and G. H. Tomlinson. Mercury - the Link Between pH of Natural Waters and the Mercury Content of Fish. Paper presented at a meeting of the Panelon Mercury of the Coordinating Committee for Scientific and Technical Assessments of Environmental Pollutants, National Academy of Sciences, National Research Council, Washington, D.C., May 3, 1977. Montreal, Quebec: Domtar Research Center, 1977.

49. Billings, W. D. Plants and the Ecosystem. Third ed. Wadsworth Publishing Company, Inc., Belmont, Pennsylvania, 1978. 177 p., pp. 1-62.

50. Odum, E. P. Fundamental of Ecology. Third ed. W. B. Saunders Co., Philadelphia, Pennsylvania, 1971. pp. 5, 8-139.

51. Kwiatkowski, R. E., and J. C. Roff. Effects of Acidity on the Phytoplankton and Primary Productivity of Selected Northern Ontario Lakes. Can. J. Bot., 54:2546-2561, 1976.

52. Hendrey, G. R., K. Baalstrud, T. S. Traaen, M. Laake, and G. Raddum. Acid Precipitation: Some Hydrobiological Changes. Ambio., 5:224-227, 1976.

53. Grahn, O. Macrophyte Succession in Swedish Lakes Caused by Deposition of Airborne Acid Substances. In: Proceedings of the First International Symposium on Acid Precipitation and the Forest Ecosystem. L. S. Cochinger and T. A. Seliga, eds. Ohio State University, May 12-15, 1975. USDA Forest Service General Technical Report NE-23, Upper Darby, Pennsylvania, Forest Service, U.S. Department of Agriculture, Northeastern Forest Experiment Station, 1976. pp. 519-530.

54. Hendrey, G. R. Effects of pH on the Growth of Periphytic Algae in Artificial Stream Channels. Research Report IR 25/76. 1432 Aas-NLH, Norway: SNSF Project Secretariat, 1976. 50 pp.

55. Grahn, O., H. Hultberg, and L. Landner. Oligotrophication--A Self-accelerating Process in Lakes Subjected to Excessive Supply of Acid Substances. Ambio., 3:93-94, 1974.

56. Conroy, N., K. Hawley, W. Keller, and C. Lafrance. Influences of the Atmosphere on Lakes in the Sudbury Area. J. Great Lakes Res., 2 (Suppl. 1):146-165, 1976.

57. Anderson, I., O Grahn, H. Hultberg, and L. Landner. Jamforande Undersokning av Olika Tekniker for Terstallande av Forsurade sjoar. STU Report 73-3651. Stockholm: Institute for Water and Air Research, 1975. Cited in: National Research Council, Sulfur Oxides, National Academy of Sciences, Washington, D.C., 1978. 118 pp.

58. Borgstrom R., J. Brittain, and A. Lillehammer. Evertebrater og Surt Vann: Oversikt over Innsamplingslokaliteter. Research Report IR 21/76. 1432 Aas-NLH, Norway: SNSF Project Secretariat, 1976. Cited in:

National Research Council, Sulfur Oxides, National Academy of Sciences, Washington, D.C., 1978. 33. pp.

58a. Sprules, G. W. Midsummer crustacean zooplankton communities in aicd-stressed lakes. J. Fish Res. Board Can. 32:389-395, 1975.

59. Oakland, J. Distribution and Ecology of the Freshwater Snails Gastropoda of Norway. Malacologia, 9:134-151, 1969.

60. Bell, H. L. Effects of Low pH on the Survival and Emergence of Aquatic Insects. Water Res., 5:313-319, 1971.

61. Pough, F. H., and R. E. Wilson. Acid Precipitation and Reproductive Success of Ambystoma Salamanders. In: Proceedings of the First International Symposium on Acid Precipitation and the Forest Ecosystem, Ohio State University, May 12-15, 1975. U.S. For. Serv. Gen. Tech. Rep. NE-23, 1976. pp. 531-544.

62. Gosner, K. L., and I. H. Black. The Effects of Acidity on the Development and Hatching of New Jersey Frogs. Ecology, 38:256-262, 1975.

63. Pough, F. H. Acid Precipitation and Embryonic Mortality of Spotted Salamanders, Ambystoma Maculatum. Science, 192:68-70, 1976.

64. U.S. Department of the Interior, Fish and Wildlife Service. Impacts of Coal-Fired Power Plants on Fish, Wildlife, and their Habitats. FWS/OBS-78/29, March 1978. p. 64-70.

65. Hendrey, G. Aquatics Task Force on Environmental Assessment of the Atikokan Power Plant: Effects on Aquatic Organisms. Land and Fresh Water Environmental Sciences Group, Department of Energy and Environment. Brookhaven National Laboratory Associated Universities, Inc., Upton, New York, 1978. 16 p.

66. Jacobsen, J. S. Experimental Studies on the Phytotoxicity of Acidic Precipitation: The United States Experience. Paper presented at NATO Advanced Research Institute, Effects of Acid Precipitation on Terrestrial Ecosystems, Toronto, May 22-26, 1978. In Press.

67. Wood, T., and F. H. Bormann. Short-Term Effects of a Simulated Acid Rain Upon the Growth and Nutrient Relations of Pinus strobus, L. Water, Air Soil Pollut., 7:479-488, 1977.

68. Evans, L. S., N. F. Gmur, and F. DaCosta. Leaf Surface and Histological Perturbations of Leaves of Phaseolus vulgaris and Helianthus annus after Exposure to Simulated Acid Rain. Amer. J. Bot., 64:093-913, 1977.

69. Hindawi, I. J., J. A. Rea, and W. L. Griffis. Response of Bush Bean Exposed to Acid Mist. 70th Annual Meeting of the J. Air Pollut. Control Assoc. Abstract 77-30.4, 1977.

70. Evans, L. S., and T. M. Curry. Differential Responses of Plant Foliage to Simulated Acid Rain. Amer. J. Bot., 66:953-962, 1979.

71. Evans, L. S., N. F. Gmur, and F. DaCosta. Foliar Response of Six Clones of Hybrid Poplar to Simulated Acid Rain. Phytopathology, 68:847-856, 1978.

72. Office of Research and Development. Research Highlights 1979. Report No. EPA-600/9-80-005, U.S. Environmental Protection Agency, 1980. pp. 26-29.

73. Evans, K. S., N. F. Gmur, and J. J. Kelsch. Perturbations of Upper Leaf Surface Structures by Simulated Acid Rain. Environ and Exptl. Bot., 17:145-149, 1977.

74. Jacobson, J. S., and P. van Leuken. Effects of Acidic Rain on Vegetation. Proc. Fourth Intern. Clean Air Congress, 1977. pp. 124-127.

75. Ferenbaugh, R. W. Effects of Simulated Acid Rain on Phaseolus vulgaris L. (Fabaceae). Amer. J. Bot., 63:283-288, 1976.

76. Abrahamsen, G., K. Bjor, R. Horntvedt, and B. Tveite. Effects of Acid Precipitation on Coniferous Forest. In: Research Report FR-6, F. H. Braekke, ed. SNSF Project, NISK, Aas, Norway, 1976. pp. 37-63.

77. Irving, P. M. Induction of Visible Injury in Chamber-Grown Soybeans Exposed to Acid Precipitation. RER Division Annual Report, ANL-78-65-III, Argonne, Illinois, 1978.

78. Wood, T., and F. H. Bormann. The Effects of an Artificial Acid Mist upon the Growth of Betula alleghaniensis. Brit. Environ. Pollut., 7:259-268, 1974.

79. Lang, D. S., D. S. Shriner, and S. V. Krupa. Injury to Vegetation Incited by Sulfuric Acid Aerosols and Acidic Rain. Paper 78-7.3, 71st Annual Meeting, Air Pollution Control Association, Houston, Texas, 1978.

80. Shriner, D. S. Atmospheric Deposition: Monitoring the Phenomenon; Studying the Effects. In: Handbook of Methodology for the Assessment of Air Pollutant Effects on Vegetation, W. W. Heck, S. V. Krupa, and S. N. Linzon, ed. Air Pollution Control Association, Pittsburgh, Pennsylvania, 1979.

81. Irving, P., and J. E. Miller. The Effects of Acid Precipitation Alone and in Combination with Sulfur Dioxide on Field-Grown Soybeans. RER Division Annual Report, ANL-78-65-III, Argonne, Illinois, 1978.

82. Shriner, D. S. Effects of Simulated Rain Acidified with Sulfuric Acid on Host-Parasite Interactions. In: Proceedings of the First International Symposium on Acid Precipitation and the Forest Ecosystem, May 12-15, 1975, Columbus, Ohio, L. S. Dochinger and T. A. Seliga, eds.

U.S. Forest Service General Technical Report NE-23, U.S. Department of Agriculture, Forest Service, Northeastern Forest Experiment Station, Upper Darby, Pennsylvania, 1976. pp. 919-925.

83. Abrahmsen, G., and G. J. Dollard. Effects of Acid Deposition on Forest Vegetation. In: Ecological Effects of Acid Precipitation, M. J. Wood, ed. Report of workshop held at Calley Hotel, Gatehouse-of-Fleet, Galloway, United Kingdom, 4-7 Sept. 1978. EPRI SOA77-403, Electric Power Res. Institute, Palo Alto, California, 1979.

84. Jonsson, B., and R. Sundberg. Has the Acidification by Atmospheric Pollution Caused a Growth Reduction in Swedish Forests? Rep. Notes No. 20, Royal College of Forestry, Stockholm, Sweden, 1972.

85. Cogbill, C. V. The Effect of Acid Precipitation on Tree Growth in Eastern North America. Water, Air and Soil Pollution, 8:89-93, 1977.

86. Abrahamsen, G., R. Horntvedt, and B. Tveite. Impacts of Acid Precipitation on Coniferous Forest Ecosystems. Water, Air and Soil Pollution, 8:57-73, 1977.

87. Tukey, H. B., Jr. The leaching of Substances from Plants. Ann. Review of Plant Physiology, 21:305-324, 1970.

88. Wood, T., and F. Bormann. Increases in Foliar Leaching by Acidification of an Artificial Mist. Ambio., 4:169-171, 1975.

89. Overrein, L. N. Sulphur Pollution Patterns Observed; Leaching of Calcium in Forest Soil Determined. Ambio., 1:145-147, 1972.

90. Anderson, R. Acidifying Effects of Nitrogen Fertilizers on Swedish Farms. Gruafor Battring, 26:11-24, 1973-1974.

91. McFee, W. W. Effects of Pollutants on Soils. In: Polluted Rain, T. Toribara, et al., eds. Pergamon Press, New York, New York, 1980. p. 307.

92. Oden, S. The Acidification of Soils due to Nitrogen Fertlization and Atmospheric Pollutant of Ammonium. Skogsand Cantvenksakad., Tidskar, 113:45-458, 1974.

93. Gorham, E. The Influence and Importance of Daily Weather Conditions in the Supply of Chloride, Sulfate, and Other Ions to Fresh Waters from Atmospheric Precipitation. Phil. Trans Roy. Soc. (London) B., 247:147-178.

94. Wiklander, L. Leaching and Acidification of Soils. In: Ecological Effects of Acid Precipitation. M. J. Wood, ed. Report of workshop held at Cally Hotel, Gatehouse-of-Fleet, Galloway, United Kingdom, 4-7 Sept. 1978. EPRI SOA77-403, Electric Power Res. Institute, Palo Alto, California, 1979.

95. Malmer, N. Acid Precipitation: Chemical Changes in the Soil. Ambio., 5:231-234, 1976.

96. McFee, W. W., J. M. Kelly, and R. H. Beck. Acid Precipitation: Effects on Soil Base pH and Base Saturation of Exchange Sites. In: Proceedings of the First International Symposium on Acid Precipitation and the Forest Ecosystem, May 12-15, 1975, Columbus, Ohio, L. S. Dochinger and T. A. Seliga, eds. U.S. Forest Service General Technical Report NE-23, U.S. Department of Agriculture, Forest Service, Northeastern Forest Experiment Station, Upper Darby, Pennsylvania, 1976. pp. 725-735.

97. Ulrich, B. Die Umweltbeeinflussung des Nahrstoffhaushaltes eines Bodensauren Buchenwalds. Forstwiss. Centralbl., 94:280-287, 1975.

98. Alexander, M. Effects of Acidity on Microorganisms and Microbial Processes in the Soil. In: Effects of Acid Precipitation on Terrestrial Ecosystems, T. C. Hutchinson and M. Havas, eds. Plenum Press, New York, New York, 1978. p 341-362.

99. Kucera, V. Effects of Sulfur Dioxide and Acid Precipitation on Metals and Anti-Rust Painted Steel. Ambio., 5:243-248, 1976.

100. Nriagn, J. Deteriorative Effects of Sulfur Pollution on Materials. In: Sulfur in the Environment Part II: Ecological Imports, J. Nriagn, ed. John Wiley and Sons, New York, 1978. pp. 482, pp.2-59.

101. Cowling, E. B., and L. S. Dochinger. The Changing Chemistry of Precipitation and its Effects on Vegetation and Materials. In: AICHE Symposium Series, Control and Dispersion of Air Pollutants: Emphasis on NO_x and Particulate Emissions, 7:134-142, 1978.

102. Sereda, P. J. Effects of Sulphur on Building Materials. In: Sulphur and its Organic Derivatives in The Canadian Environment, Nat. Res. Council of Canada. NRC Associate Committee on Scientific Criteria for Environmental Quality, Ottawa, Canada, 1977. p. 359-426.

103. Glass, G. E., and O. L. Loucks, eds. Impacts of Air Pollutants on Wilderness Areas of Northern Minnesota. Environmental Research Laboratory - Duluth, ORD, U.S. Environmental Protection Agency, Duluth, Minnesota, 1979. p. 188.

104. Comments of the American Petroleum Institute on the Revised Air Quality Criteria For Oxides of Nitrogen. Appendix C titled, "Reinterpretation of Data Used to Evaluate Effects of Acidic Precipitation Upon Fish Stocks in Pennsylvania Streams", November 1980.

Section 5

Mitigative Strategies

EMISSION REDUCTIONS

It is generally acknowledged that most man-made emissions of acid rain precursors in the United States are from powerplants, industrial combustion, transportation, and non-ferrous smelters. Since smelters are quite remote from the sensitive areas in the northeastern U.S., attention has been placed on the remaining three source categories. A broad assessment utilizing the Strategic Environmental Assessment System (SEAS) is now underway, but results have not been generated. A more narrowly focused analysis of electric powerplants has produced preliminary results. A summary of potential regulatory options to control acid rain is contained in Table 5-1.

The powerplant effort is co-funded by DOE and EPA; it utilizes two contractors, ICF, Incorporated, and Teknekron Research, Incorporated. Only results for the ICF study are available at this time. They have not undergone extensive review by the sponsors, so the results must be considered preliminary in nature. To date only SO_2 related strategies have been assessed.

The scenarios evaluated generally obtained a 10-38% reduction in SO_2 emissions for roughly $1-4 billion per year (annualized costs). They have minimal impact on overall coal demand, but do tend to disrupt traditional high sulfur coal markets in Northern Appalachia and the Midwest.

REGULATORY ALTERNATIVES EVALUATED

At the direction of EPA and DOE several regulatory alternatives were identified for evaluation. These included alternatives to reduce utility SO_2 emissions and combined NO_x and SO_2 emissions. The alternatives for which interim results are presented are:

- Base Case -- no new Acid Rain initiatives; no change in the current environmental regulations.

- 10 Percent Broad Regional Rollback -- a 10 percent rollback in SO_2 emissions projected in the Base Case for the 31-state Acid Rain Region (see Figure 1). This was treated as a broad regional emission limit which permitted trading of emission rights among different utilities and facilities in different states within this region.

TABLE 5-1. SUMMARY OF POTENTIAL REGULATORY OPTIONS TO CONTROL ACID RAIN

Strategy	Advantages	Disadvantages
Enforcement of current SIPS	Major reductions in emissions are possible.	Monitoring and enforcement in terms of personnel and equipment are very great.
Assign emission caps on a state or regional basis (in lb/MMBtu)	Uniform, simplistic, ease of administration, emission reductions would be achieved.	High sulfur coal states might have to devise method to protect local miners (like in NSPS).
Create economic incentives for SO_2 and/or NO_x control through emission taxes or marketable permits	State equity would be maintained; emission reductions would be achieved.	Implementation and monitoring of this approach would be cumbersome.
Wait until LIMB (Limestone Injection with Modified Burners) is available (EST 1986-1988) to get control of both SO_x and NO_x	High SO_x reduction (60-70%) as well as high NO_x reduction at costs *well below* those for conventional scrubbers. Major uncertainties regarding the acid rain phenomenon might be resolved as this technology becomes commercially available.	Long lag time before technology is commercially available.
Assign regional pollution reduction levels	Less costly than most other options.	No state discretion; equity issue would be raised; administrative difficulties.
Specific percent reduction in each state's emissions	Administratively easy to apply.	Environmentally progressive states would be penalized in that even further emissions reductions would be required; more costly than previous option (regional reduction levels).
Utilize Section 115 of the Clean Air Act which sets up an international agreement to control acid rain	Plenty of state flexibility allowed in the revision of individual state implementation plans.	Equity issue would be raised; administrative difficulties.

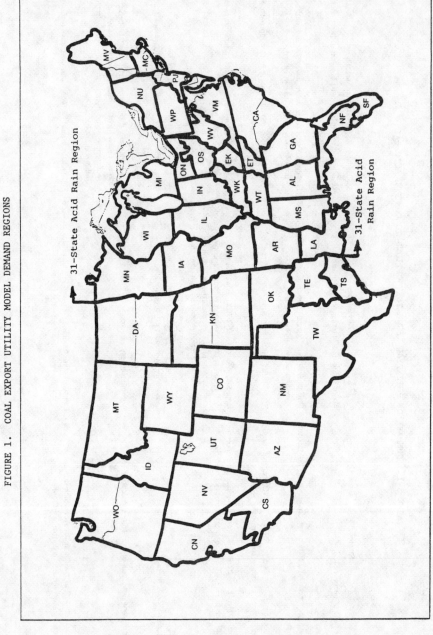

FIGURE 1. COAL EXPORT UTILITY MODEL DEMAND REGIONS

- 30 Percent Broad Regional Rollback -- a 30 percent rollback in SO_2 emissions projected in the Base Case for the 31-state Acid Rain Region.

- 10 Percent Small Regional Rollback -- a 10 percent rollback in SO_2 emissions projected in the Base Case for each of the ICF utility demand regions (45 regions in all -- see Figure 1). This was treated as an emission limit which permitted trading of emission rights among utilities and facilities only within each of the individual demand regions.

- 30 Percent Small Regional Rollback -- a 30 percent rollback in SO_2 emissions projected in the Base Case for each of the ICF utility demand regions.

- 4.0 Pound Emission Limit -- the current emission limits specified in applicable State Implementation Plans (SIP's) were restricted to a maximum of 4.0 pounds SO_2 per million Btu.

- 2.0 Pound Emission Limit -- the current emission limits specified in applicable State Implementation Plans (SIP's) were restricted to a maximum of 2.0 pounds SO_2 per million Btu.

Forecasts of utility emissions and costs and coal production were developed for the years 1985, 1990, 1995, and 2000. The Base Case also included a forecast for 2010 which is not presented.

Further, several of the above alternative regulations were evaluated with the additional stipulation that some protection would be provided for possible short-term impacts on regional coal production and mine employment. This "protection" was modeled to require that for any forecast year, the regional coal production occurring in 1985 in the Base Case would at a minimum be achieved.

For the "local coal" requirements, the following regulatory alternatives were evaluated:

- 30% Broad Regional Rollback -- a 30 percent rollback in SO_2 emissions in the Base Case for the 31-state Acid Rain Region.

- 30% Small Regional Rollback -- a 30 percent rollback in SO_2 emissions in the Base Case for each of the ICF utility demand regions.

- 4.0 Pound Emission Limit -- the current emission limits specified in applicable SIP's were restricted to a maximum of 4.0 pounds SO_2 per million Btu.

- 2.0 Pound Emission Limit -- the current emission limits specified in applicable SIP's were restricted to a maximum of 2.0 pounds SO_2 per million Btu.

The results for each of the forecast years (1985, 1990, 1995, and 2000) for all of the above regulatory alternatives are summarized in the following four tables. Note that these summary results provide national aggregate figures for comparison purposes. It should be recognized that very important regional differences in utility emissions and costs and coal production result from the various regulatory alternatives. These regional effects are not presented here, but will be identified in the draft report from ICF to be issued at a later date.

LIMING

Another possible option for use in a program for the management of acid deposition may be to mitigate its harmful effects in susceptible areas. If such an option proved practical, it might be used either alone or in conjunction with emission control strategies. Methods suggested have included increasing the pH of affected lakes, soils, forests, etc.; the development of protective coatings for exposed structures and materials; and the development of acid-resistant species of crops, trees, and fish. Only the first approach, which involves liming of lakes and/or streams, has received any investigation thus far.

The term "liming" applies to any procedure whereby the pH of an acidified lake is raised. Various substances have been used to achieve this purpose including soda ash, potash, dolomite, calcium hydroxide, calcium oxide, and limestone. Limestone application has proved to be the cheapest.

Just as in the agronomic application of limestone, commonly done to counteract the effects of acidification from fertilizers, the financial cost of the liming treatment would have to be carefully balanced against the loss incurred if the treatment were not given. This balance could be a major consideration in remote or generally inaccessible areas.

Liming of acidified lakes was first attempted in Sweden and effects have been well documented. Henriksen and Johannessen[1] have presented a survey of the literature. Swedish studies have also been reported by Grahn and Hultberg[2,3] and Hultberg and Grahn.[4] Wright[5] has reviewed an early attempt in which addition of chalk to Swedish lakes increased pH and led to increased phytoplankton growth and improved fish survival. The reclamation of acidified lakes in Canada has been discussed by Scheider et al.[6]

The New York State Bureau of Fisheries has added lime to 51 ponds since the mid-1950s. The emphasis was on maintaining fish populations that were unique or especially important for recreational activities. However, as mentioned, the logistical difficulties encountered in attempting to transport large quantities of limestone to isolated Adirondack ponds often made the final benefits realized somewhat questionable. Further studies are required to assess the true feasibility of this approach and the methodologies that would need to be applied.

1990 FORECASTS—INCREMENTAL CHANGES FROM BASE CASE

	Base	#2C 10% ARM	#3C 30% ARM	#4C 10% Each	#5C 30% Each	#6C 4 Pound	#7C 2 Pound
SO_2 Emissions (10^6 tons/yr.)							
ARM	17.1	-1.7	-5.1	-1.7	-5.1	-2.7	-7.0
Other	1.8	-	-	-0.2	-0.5	-	-0.2
Total	18.9	-1.7	-5.1	-1.9	-5.7	-2.7	-7.1
NO_x Emissons (10^6 tons/yr.)							
ARM	6.7	-	-0.1	-	-0.1	-	-
Other	2.3	-	-	-	-	-	-
Total	9.0	-	-0.1	-	-0.1	-	-
Annualized Costs ($ 1980 x 10^9/yr.)							
ARM	70.7	+0.1	+0.9	+0.4	+1.4	+0.7	+2.2
Other	39.9	-	-	+0.9	+2.5	-	+0.4
Total	110.6	+0.1	+0.9	+1.2	+3.9	+0.8	+2.7
Annualized Costs (% Change in Rate)							
ARM	-	+0.1	+0.7	+0.3	+1.1	+0.6	+1.8
Other	-	-	-	+1.4	+3.9	-	+0.7
Total	-	+0.1	+0.5	+0.7	+2.0	+0.4	+1.4
Dollars/Ton SO_2 Removal (1980 $)[1]							
ARM	-	-	-	-	-	-	-
Other	-	-	-	-	-	-	-
Total	-	+50	+170	+210	+470	+280	+480
Additional Scrubbed Capacity (Gw)[2]							
ARM							
Wet	27.1	+0.3	+7.3	+0.6	+8.6	+12.4	+25.8
Dry	16.2	+0.5	+0.6	+0.8	+4.2	-1.8	+8.6
Other							
Wet	23.6	-	-	-0.8	-0.5	+1.0	+3.8
Dry	13.8	+0.2	+0.2	+2.9	+15.2	-1.0	-
Total							
Wet	50.6	+0.3	+7.4	-0.2	+8.0	+13.4	+29.6
Dry	30.0	+0.7	+0.8	+3.7	+19.4	-2.8	+8.7

[1]/ This measure of the costs to remove SO_2 includes a measure of both the costs to utilities (reflected by the annualized costs above) and the changes in delivered coal costs to non-utility coal consumers. Regional estimates of this measure will be provided in a later ICF draft report.

[2]/ Scrubbed capacity means the total Gw of the <u>units</u> on which scrubbers are built.

1990 FORECASTS INCREMENTAL CHANGE FROM BASE CASE

	Base	#2C 10% ARM	#3C 30% ARM	#4C 10% Each	#5C 30% Each	#6C 4 Pound	#7C 2 Pound
Utility Fuel Consumption (Quads)							
Coal	16.8	—	—	—	—	—	—
Oil	2.1	—	—	—	—	—	+0.1
Gas	3.9	—	—	—	—	—	—
Nuclear	7.8	—	—	—	—	—	—
Other	3.4	—	—	—	—	—	—
Total	34.0	—	—	—	—	—	+0.1
Coal Production (10⁶ tons)							
Northern Appalachia	209	-1	-20	-5	-18	-3	-23
Central Appalachia	318	+6	+28	+5	+23	+4	+34
Southern Appalachia	22	—	—	—	—	—	+2
Midwest	174	-15	-25	-10	-28	-18	-44
Central West	18	—	+1	—	+2	—	+1
Gulf	57	+2	+2	+3	+2	—	—
Eastern Northern Great Plains	26	—	—	—	—	—	—
Western Northern Great Plains	260	-2	+4	-6	-8	+4	+5
Rocky Mountains	81	+2	+2	+3	+6	+6	+15
Southwest	71	+5	+7	+6	+17	+5	+8
Northwest	16	+3	+3	+3	+3	+1	+1
Alaska	—	—	—	—	—	—	—
Total	1,252	—	+1	—	-3	—	-2
Western Coal to East	81	+8	+17	+8	+20	+10	+22

Mitigative Strategies   195

LOCAL COAL RUNS

1990 FORECASTS -- INCREMENTAL CHANGES FROM BASE CASE

	Base	#1 LC 30% ARM	#2 LC 30% Each	#3 LC 4 Pound	#4 LC 2 Pound
SO_2 Emissions (10^6 tons/yr.)					
ARM	17.1	-5.1	-5.1	-2.7	-7.0
Other	1.8	-	-0.6	-	-0.2
Total	18.9	-5.1	-5.7	-2.7	-7.2
NO_x Emissions (10^6 tons/yr.)					
ARM	6.7	-0.1	-0.1	-	-
Other	2.3	-	-	-	-
Total	9.0	-0.1	-0.1	-	-
Annualized Costs (\$ 1980 x 10^9/yr.)					
ARM	70.7	+1.2	+1.5	+1.0	+2.3
Other	39.9	+0.1	+2.6	-	+0.4
Total	110.6	+1.2	+4.1	+1.0	+2.7
Annualized Costs (% Change in Rate)					
ARM	-	+1.0	+1.2	+0.9	+1.9
Other	-	+0.1	+4.0	-	+0.7
Total	-	+0.7	+2.3	+0.6	+1.3
Dollars/Ton SO_2 Removal (1980 \$) [1]					
ARM	-	-	-	-	-
Other	-	-	-	-	-
Total	-	210	490	480	290
Additional Scrubbed Capacity (Gw) [2]					
ARM					
Wet	27.1	+9.1	+11.6	+13.3	+31.3
Dry	16.2	+0.6	+4.2	-1.4	+5.1
Other					
Wet	23.6	-	-0.5	+0.8	+4.1
Dry	13.8	+0.2	+15.9	-0.8	-0.2
Total					
Wet	50.6	+9.1	+11.1	+14.1	+35.4
Dry	30.0	+0.8	+20.1	-2.2	+4.9

[1] This measure of the costs to remove SO_2 includes a measure of both the costs to utilities (reflected by the annualized costs above) and the changes in delivered coal costs to non-utility coal consumers. Regional estimates of this measure will be provided in a later ICF draft report.

[2] Scrubbed capacity means the total Gw of the *units* on which scrubbers are built.

LOCAL COAL RUNS
1990 FORECASTS -- INCREMENTAL CHANGES FROM BASE CASE

	Base	#1 LC 30% ARM	#2 LC 30% Each	#3 LC 4 Pound	#4 LC 2 Pound
Utility Fuel Consumption (quads)					
Coal	16.8	--	--	--	--
Oil	2.1	--	--	--	--
Gas	3.9	--	--	--	--
Nuclear	7.8	--	--	--	--
Other	3.4	--	--	--	--
Total	34.0	--	--	--	+0.1
Coal Production (10⁶ tons)					
Northern Appalachia	209	-11	-11	-1	-19
Central Appalachia	318	+13	+13	+2	+24
Southern Appalachia	22	+2	+2	+2	+2
Midwest	174	-21	-25	-15	-26
Central West	18	+1	+1	+1	+1
Gulf	57	+2	+2	--	--
Eastern Northern Great Plains	26	--	--	--	--
Western Northern Great Plains	260	+9	-2	+3	+4
Rocky Mountains	81	-2	-1	+5	+9
Southwest	71	+7	+16	+4	+4
Northwest	16	+3	+2	--	+1
Alaska	0	--	--	--	--
Total	1,252	+5	-1	+1	+1
Western Coal to East	81	+22	+20	+8	+15

Henriksen and Johannessen s report details the forms of calcium that could be used and options available for applications. They also present methods of calculating the lime required to raise the pH of a given body of water to 5.5.

Various factors must be considered in attempting to assess the feasibility of liming acidified waters. The efficiency of liming in terms of maintenance of elevated pH requires careful consideration of lake turnover times and the sources of acid input to the lake. The addition of lime to streams has been carried out for many years and often has advantages over the direct application of lime to lakes, although this has been successful, as well. Other systems for practical application of lime include lime wells that consist of a column of lime in a river bank. Water diverted to the base of this column and subsequently entering a lake offsets any pH reduction occurring with spring snow melt while avoiding potential problems associated with sporadic addition of large amounts of lime to treated waters.

In addition, expected and resulting chemical, biological, and ecological changes occurring in lake waters must be carefully studied subsequent to liming. Although to date, no observations of long-term detrimental effects of liming have been observed, the true ecological consequences are unknown. Invariably, alkalinity and pH will increase. Phosphorous release from lake sediments should be observed, probably as a result of ion-exchange processes with HCO_3^- generated from liming. Concentrations of Zn, Mn, and Al drop with elevations in pH.

Among the biological changes that have been reported subsequent to liming are increased phytoplankton and zooplankton diversity. In fish populations whose age distribution was skewed toward older groups as a result of acidification, liming resulted in restoration of younger age groups. Increases in total fish biomass have been observed. Addition of $CaCO_3$ and $Ca(OH)_2$ to two acidic lakes in Sudbury, Ontario, increased pH, decreased heavy metal concentrations, and caused a temporary decline in chlorophyll.[6,7]

There are potential problems associated with liming of waters to counteract acidification and data are lacking in these areas. Emmelin[8] has reported that the concentrations of toxic metals in some commercial limestones may pose potential concern. Dickson[9] has reported that aluminum leached from the soil by acid precipitation is especially hazardous to fish populations after liming and several fish kills can be attributed to this cause. Aluminum toxicity is apparently increased at pH levels between 4 and 6. He also states that liming increases cadmium accumulation in lakes. Research obviously is needed to quantify these problems before general liming programs can be initiated.

As mentioned, treatment of affected waters by large-scale liming programs would represent major undertakings that could be logistically difficult and expensive. Detailed evaluation of the economic realities involved in such programs must be made. Henriksen and Johannessen[1] have estimated costs to be a minimum of 6-20 million Pounds Sterling per year

198   Acid Rain Information Book

at 1978 prices. This figure only includes wet deposition and does not, perhaps, reflect the true labor costs that would be required if wide-scale liming of whole lakes was initiated. R. A. Barnes in a recent Air Pollution Control Association Acid Rain Panel discussion estimated the cost of liming affected areas in Europe to be $150 million, which he believed was a reasonable cost alternative when compared to the costs of controlling combustion emissions.[11]

An example of the complexities involved in undertaking a lake-liming program is illustated in Figure 5-1. This proposed lake restorative method is called <u>Contracid</u> and has recently been reported in the Swedish press.[10] As shown in Figure 5-1, this method would involve a physical "raking" of the bottom of the lake in a manner similar to harrowing farm land. Appropriate chemical mixtures to raise the lake sediment's pH (sodium hydroxide or sodium carbonate) would be injected using a compressed air dosage system. Theoretically, exchanging hydrogen cations for sodium cations would serve to neutralize the acidity of hydrogen ions.

REVIVING A LAKE:

1. Field lab
2. Chemicals supply
3. Portable compressor
4. Chemical dilution tank
5. Harrow
6 and 7. Air feed lines
8. Air driven pump
9. Mixed chemicals supply line
10. Guide line
11. Air driven dilution pump
12. Dilution water intake
13. Pneumatic tugging arrangement.

Figure 5-1. <u>Contracid</u> method of lake restoration.[10]

The Contracid method has been funded to perform a full-scale experimental restoration of acidified Lake Lilla Galtsjon in southern Sweden this September. Follow-up studies will continue through snowmelt in the spring of 1981. Liming strategies have not reached this stage in the United States and it will be important to follow results closely.

REFERENCES

1. Henriksen, A. and M. Johannessen. Deacidification of Acid Water. SNSF Project, Report IR 5/75 Aas, Norway, 1975.

2. Grahn, O. and H. Hultberg. Development of Methods for Liming Acidic Running Waters. Swedish Water and Air Pollution Research Laboratory Report, 1975.

3. Grahn, O. and H. Hultberg. Importance of Grain Size and Geological Origin for Neutralization Effectiveness of Three Different Calcium Carbonates. Swedish Water and Air Pollution Research Laboratory Report, 1976.

4. Hultberg, H. and O. Grahn. Some Effects of Adding Lime to Lakes in W. Sweden. Swedish Water and Air Pollution Research Laboratory Report, 1975.

5. Wright, R. F. Acid Precipitation and its Effects on Freshwater Ecosystems: An Annotated Bibliography. In: Proceedings of First International Symposium on Acid Precipitation and the Forest Ecosystem, May 12-15, 1975, Columbus, Ohio, L. S. Dochinger and T. A. Seliga, eds. U.S. Forest Service General Technical Report NE-23, U.S. Department of Agriculture, Forest Service, Northeastern Forest Experiment Station, Upper Darby, Pennsylvania, 1976. pp. 619-678.

6. Scheider, W., et al. Reclamation of Acidified Lakes near Sudbury, Ontario. Ontario Ministry of the Environment Report, Rexdale Ontario, 1975.

7. Michalski, M. F. P. and J. Adamski, Restoration of Acidified Lakes Middle and Lohi in the Sudbury Area. In: Proceedings of the Ontario Industrial Wastes Conference, 1974. pp. 163-175.

8. Emmehn, L. Environmental Planning in Sweden--the State of Sweden's Lakes. Current Sweden, 79:1, 1977.

9. Dickson, W. Experience from Small Scale Liming in Sweden. In: Proc. Int. Symp. on Sulfur Emissions and the Environment, London, May 1979.

10. Anonymous. A Harrowing Experience! Reviving Acidified Lakes. Sweden Now, 14 (2): 38, 1980.

11. Barnes, R. A. Acid Rain - An International Concern. An APCA Panel Presentation. J. Air Poll. Cont. Assoc. 30: 1089-1097, 1980.

12. Bengtsson, B., W. Dickson, and P. Nyberg. Liming Acid Lakes in Sweden, Ambio, 9(1): 34, 1980.

13. The Swedish Freshwater Fisheries Laboratory, Report No. 8, 1979. The Liming of Lakes and Waterways, 1977-1979.

14. Dickson, W. Some Effects of the Acidification of Swedish Lakes. Verh. Internat. Verein. Limnol., 20: 851, 1978.

15. Fleischer, S. and L. Halmstad. The River Hogvadsan Liming Project - A Presentation Internat. Conf. on the Ecological Impact of Acid Precipitation, Sandefjord, Norway, March 1980.

Section 6

Summary of Issues, Uncertainties and Further Research Needs

This section summarizes the key issues, uncertainties, and further research needs relating to what is known of and speculated about acid rain. Table 6-1, at the end of this section, is a condensed summary of the individual issues that have surfaced from a comprehensive review of the literature. These issues, which are discussed in detail throughout the book, are organized according to major subject areas, covering sources, atmospheric chemistry and physics, monitoring effects, and mitigating strategies. Corresponding to each issue is an indication of the level of uncertainty (viz., low, moderate, or high) as suggested in the literature. Although sometimes speculated, the level of uncertainty is often directly inferred from the literature by a lack of concensus among the experts on how a given issue relates to acid rain. In other cases, there is a clear gap in the data that prevents cause and effect relationships from being established. Finally, examination of the issues has often led to identification of specific research needs and level of intensity that may be required to alleviate the concern or uncertainty underlying an issue. These research needs are identified, where possible, to help serve as the basis in developing mitigative strategies.

In view of these issues and their stated levels of uncertainty, research groups and steering committees have outlined several major or aggregate issues to be addressed so that data gaps in the present knowledge on the cause and effect relationships concerning the acid rain phenomenon can be filled. The total resources required for these efforts are difficult to estimate because of the complexity and multidisciplinary talents needed to comprehensively address each issue. One of the most important of these factors is that the character of anthropogenic emissions will be constantly changing as future energy scenarios, pollution abatement procedures, mandated control requirements, and industrial processes are implemented.

The recently revised Federal Acid Rain Assessment Plan (August 1980),[1] prepared by the President's Acid Rain Coordination Committee, contains a comprehensive strategy to expand present understanding of the phenomenon and effects of acid deposition. Recommendations from that plan are presented below.

- A permanent, nationwide monitoring network (National Trends Network) for acid deposition should be established to provide a continuous high-quality record of temporal and spatial trends in the chemistry of wet and dry deposition in all major regions of the United States.

- Additional research should be initiated at once, especially in the following critical areas:

    - investigation of the effects of acid precipitation on the well-being and productivity of crops, forests, soils, and aquatic ecosystems;

    - development of predictive models or simple measurements to determine the vulnerability of lakes, streams, soils, and materials to continuing acid deposition;

    - development of methods for reliable measuring or estimating dry deposition; and

    - development of a predictive capability to determine how changes in the spatial and temporal emission patterns affect the spatial, temporal, and chemical patterns of deposition in various regions of the United States.

- Existing information on the causes and consequences of acid deposition should be utilized fully in formulating interim approaches to the control of acid deposition and/or mitigation of its effects.

Several areas for future work have been identified in the "Proceedings of the Advisory Workshop to Identify Research Needs on the Formation of Acid Precipitation."[2]

- Instrumentation: The first step should consist of accurately sampling and determining the chemical composition and controlling physical parameters involved in the formation of cloud water over temporal and spatial scales. This sampling should be supplemented with continuous aerosol monitoring.

- Field Studies: A comprehensive nationwide precipitation chemistry network is needed to document the current and future chemical composition of precipitation. Field data must be gathered for 10 key parameters controlling cloud and precipitation scavenging efficiencies and cloud chemistry.

- Modeling: Atmospheric models should incorporate heterogeneous chemical and physical processes and key variables identified in field studies.

- Laboratory Studies: Determination of the kinetics and intermediate products of sulfur and nitrogen transformation in the atmosphere is required.

- Data Analysis and Interpretation: An improved emission inventory of natural and anthropogenic sources of sulfur and nitrogen compounds is required. Emphasis should be placed on rectifying discrepancies that occur in current emission inventories when a fine grid level of analysis is used. 'This information must be analyzed and correlated with accurate precipitation chemistry data from field studies to show trends and variability over temporal and spatial scales.

The issues discussed in Table 6-1 and those underlying the recommendations of the two study groups described above, demonstrate that insufficient knowledge exists regarding the explicit causes and total effects of acid rain. Research efforts by several groups are planned and underway to resolve basic issues and uncertainties. Details of these efforts are described in Section 7.

TABLE 6-1. SUMMARY OF ISSUES, UNCERTAINTIES, AND FURTHER RESEARCH NEEDS

Item/Issue	Level of Uncertainty Sources	Need for Further Research (Intensity Indicated Where Possible)
Magnitude assessment of naturally produced SO_x relative to manmade SO_x emissions (global).	*Moderate*: Natural SO_x emission estimates have been constantly revised. The latest reported estimate attributes about two-thirds of all sulfur emissions to man's activities.	*Low-Moderate*: Even though some natural sources of SO_x may be significant, they are globally distributed, whereas manmade emissions are much more concentrated. Natural sources may be responsible for background acidity in remote areas.
Magnitude assessment of naturally produced SO_x relative to manmade SO_x emissions (regional, e.g., eastern U.S., eastern Europe).	*High*: In the eastern U.S., manmade emissions account for over 90% of total SO_x (based on a sulfur budget). Natural emissions account for 4%, with inflow to the region making up the remainder. Natural emissions of SO_x in eastern Europe have been estimated at 10% using sulfur budgets.	*Low-Moderate*: (See SO_x global.) In polluted urban airsheds, it appears that anthropogenic sources of SO_x predominate. Extensive source control may not significantly alter the acidity of rain.
Magnitude assessment of naturally produced NO_x relative to manmade NO_x emissions (global and regional).	*Moderate-High*: Global fluxes of nitrogen compounds are based largely on extrapolation of experimentally determined small-scale emission factors to global scale, or the use of mass balances to obtain crude estimates for unknown sources. The ratio of natural sources to anthropogenic sources has been estimated to be as low as 1:1 to as high as 15:1. Emission inventories on a regional basis are not well defined.	*Low-Moderate*: In spite of the uncertainty cited for natural production of NO_x, these emissions are globally distributed, whereas anthropogenic emissions are much more concentrated. In polluted urban airsheds, it appears that manmade emissions are much more concentrated.
Magnitude assessment of anthropogenically produced SO_x and NO_x (EPA regional emission inventory basis.)	*Low*: There is fairly good agreement among different inventories for SO_x and NO_x on an EPA regional basis (viz., within 20%, often within 10%).	*None*: Some benefit will be realized when improvements are made in data collected at the state, AQCR, county, and facility level.
Magnitude assessment of anthropogenically produced SO_x and NO_x (state emission inventory basis).	*Moderate*: Comparison of emission inventories on a state basis begins to reveal point and area source discrepancies. Quality of state emission inventories can vary widely from state to state.	*Moderate*: Emission inventories should continue to be developed. Data reporting procedures and accuracy of data supplied by states should also be improved.
Magnitude assessment of anthropogenically produced SO_x and NO_x (AQCR, county, or facility basis).	*High*: Emission factors, methods of fuel allocation for area and mobile sources, and distribution of source types will affect accuracy of emission inventory estimates to a large degree. Some large discrepancies have been noted among different data bases.	*High*: Improvements in emission estimates and maintenance of current data in data files are required.
Projections for SO_x and NO_x to year 2000 based on the Second National Energy Plan (NEP II).	*Moderate*: Projections will depend in large part on the energy scenarios selected.	*Moderate*: Scenarios based on new control technology measures and requirements should be continually updated.
Effect of combustion variables on source emissions.	*High*: For older coal-fired power plants on prolonged retirement schedules, control of combustion variables may be more important than that for newer, more efficiently optimized combustion units.	*High*: Research has been funded and is ongoing.

(continued)

TABLE 6-1 (continued).

Item/Issue	Level of Uncertainty	Need for Further Research (Intensity Indicated Where Possible)
Oil-fired burners as direct sources of SO_3, SO_4, and H_2SO_4.	*High*: Higher SO_3 emissions result from oil-fired units than from coal-fired units, for a given amount of fuel sulfur. Emission of significant quantities of sulfates is also common from fuel oil combustion (possibly increasing emission to and formation of H_2SO_4 in the atmosphere) probably due to catalytic oxidation of trace metals in fuel oils.	*High*: Both SO_3 and H_2SO_4 are acidic themselves and may impact significantly on acid rain formation. The capital scale of impacts requires investigation.
Effect of control technology on source emissions.	*High*: Based on recently promulgated New Source Performance Standards, strict control levels must be maintained, viz., 0.6 lb/10^6 Btu for NO_x (0.7 lb/10^6 Btu for lignite-fired cyclone boiler) and 90% control of SO_x where the uncontrolled emissions are >0.6 lb/10^6 Btu but <1.2 lb/10^6 Btu.	*High*: Continued research on the feasibility and cost effectiveness of conventional and novel control measures for SO_x and NO_x is recommended, particularly on older relatively uncontrolled plants.
Effect of seasonal fuel use on source emissions.	*Low-Moderate*: On a nationwide basis, the variation in NO_x emissions from fossil-fuel generating plants is estimated at 15%, with production being greatest in the summer and least in the spring. Substantial seasonal and diurnal variations associated with mobile sources have been noted.	*Moderate*: Data bases should be continually updated. For example, reporting mobile sources by annual emissions data may underestimate their potential for producing short-term (peak) concentrations.
Contribution of ammonia to acid precipitation.	*Moderate*: Ammonia is a known alkaline substance with acid neutralizing properties. There are large amounts of ammonia in the atmospheric background relative to the acidic forms into which it can be transformed. To the extent that neutralization occurs, the free acidity of rainfall will be reduced.	*Moderate*: A low level of effort should be conducted to investigate the effects of ammonia production facilities and natural sources on local environmental conditions.
Contribution of chloride to acid precipitation.	*Moderate*: Chloride contribution in acid rain is suspected to result more from chloride releases from the burning of coal than from marine salt spray.	*Low*: A low level of effort should be conducted to ascertain local effects.
The role of ozone in sulfate and nitrate formation.	*Moderate-High*: Chemical reactions leading to the formation of acid rain precursors have been associated with gas phase and heterogeneous (gas-liquid and gas-solid) processes related to photochemical smog. Also, the scavenging of ozone by NO in plumes from major fuel-burning installations may lead to further nitric acid production.	*Moderate-High*: More research is needed to assess the synergistic role of ozone and associated photochemical products in the formation of sulfates and nitrates.

TABLE 6-1 (continued).

Item/Issue	Level of Uncertainty	Need for Further Research (Intensity Indicated Where Possible)
Contributions of hydrocarbons to acid precipitation.	*Moderate*: Hydrocarbons interact with ozone and NO_x in photochemical smog which may be a precursor condition for acid rain formation and precipitation	*Moderate*: More research is needed to assess the synergistic role of hydrocarbons in localized acid rain formation and deposition.
The role of carbon dioxide in acid rain formation.	*Moderate*: The background pH, produced by CO_2 in equilibrium with water (carbonic acid), can also be effected by other gases and soluble particles. Below a pH of 5, carbonic acid has no further acidifying (synergistic effect on rain water).	*None*.
The role of alkaline airborne dust in neutralizing acid rain.	*Moderate-High*: Alkaline dust from natural and anthropogenic sources may react with and neutralize acids in the atmosphere. The lower pH (6 to 7) of dust generated from eastern soils may render this area more susceptible to acid precipitation, but th s may be offset by the higher pH (9 to 11) of fly ash emissions.	*Moderate-High*: More research is needed to determine whether residence times of dusts are sufficient to accomplish significant neutralization of acid precipitation.
The role of airborne particulates in catalytic oxidation of SO_x to sulfates.	*High*: Catalytic oxidation of SO_x by suspended fly ash may occur in power plant plumes. Metallic constituents of the fly ash are suspected to act as catalysts. It has been noted that oil-fired boilers can em it higher concentrations of metallic catalysts than coal-fired boilers. The conversion of SO_x to particulate sulfates increases atmospheric lifetime and facilitates long-range transport.	*High*: More research is needed to define the role of metallic constituents in the catalytic oxidation of SO_x to sulfates.
.... Atmospheric Chemistry and Physics		
Deposition rates.	*High*: Knowledge of wet and dry deposition rates as a function of chemical species, air/cloud concentrations, meteorological conditions, and surface parameters is limited.	Generalizations based on continuing measurement programs are required. Moderate effort over long term is indicated. Validated dry deposition data is lacking.
Transformation rates.	*High*: Knowledge of NO_x transformation rates is becoming important because of increasing contributions of nitrates to acid deposition. Current knowledge is fragmentary. SO_x transformation rates are better understood, but uncertainties still exist. Role of other pollutants, such as trace metals, organic acids, ozone, and basic dust needs clarification.	Better understanding of chemistry required as a function of pollutant concentrations, ambient conditions, etc. Both laboratory and atmospheric studies are needed. Atmospheric studies involving airborne sampling platforms are expensive and are best done intensively over a relatively short period of time.

TABLE 6-1. (continued)

Item/Issue	Level of Uncertainty	Need for Further Research (Intensity Indicated Where Possible)
Topographic influences.	*Low*: Terrain can influence rainfall patterns and, in turn wet removal rates. May be important in certain areas.	Analysis of current monitoring, meteorological and emissions data should be used to judge importance in U.S. and Canada. Special field studies in selected areas could then be conducted.
Effect of tall stacks on long-range transport.	*Moderate*: Increased plume height reduces local surface scavenging, increases amount of pollutant leaving near-stack area, and increases time for chemical transformation before deposition.	*Low*: Better estimates of effect will come from improved understanding of deposition and transformation rates. Nocturnal emissions released above an inversion layer may be important.
Regional modeling.	*High*: Models represent the link between sources and deposition and are the only means of assessing in advance the effects of changing conditions. Current regional models suitable for evaluating the acid rain problem are preliminary and involve tentative assumptions and major simplifications, including: (1) neglect of NO_x transformation; (2) linear SO_x transformation rates; (3) elementary consideration of in-cloud chemistry; (4) the use of poorly documented decay rates for wet removal of sulfates and nitrates; (5) prescription of regional wind and precipitation fields from existing networks; and (6) representation of atmosphere by one well-mixed vertical layer.	Additional development, testing, and validation will be required before models can be used with confidence. This is an iterative process and improvements occur gradually as a result of more realistic assumptions and more detailed monitoring data.
	Monitoring	
Measurement of acidic precipitation in North America (<5.6 pH).	*Moderate-High*: Acidic precipitation has been documented in areas of North America including northeastern United States, Florida, Minnesota, Colorado, California, and Canada. Many areas in between have not yet been monitored for acidity.	*High*: Monitoring networks have been set up in the United States and Canada. Some measure acidity by event, but most take weekly or monthly samples. More work needs to be done to distinguish between wet and dry deposition.
Evidence of trends toward increasingly acidic rain over increasing area of influence.	*High*: Not enough data have been collected in the United States and Canada to firmly document a trend toward increasing acidity or to indicate that the geographic area affected is spreading.	*High*: Early methods of analysis did not directly measure acidity. Modern networks maintained over a minimum time period of about 10 years may be necessary to draw firm conclusions.
Relative importance of acidic components in rainfall.	*Moderate-High*: Acidic rainfall is generally considered to be composed of 60% sulfates, 30% nitrates, and 10% chlorides and other acidic components. These proportions vary among regions and with time and are important when determining sources. The relative importance of nitrates seems to be increasing.	*Moderate*: The ionic content of precipitation should be monitored over a period of time for all regions. This is currently being undertaken.

(continued)

TABLE 6-1 (continued).

Item/Issue	Level of Uncertainty	Need for Further Research (Intensity Indicated Where Possible)
Determination of sources of acidic components of precipitation.	*High*: The many links between anthropogenic pollution emissions and measured acidity of rainfall are unclear. The current methods include association with fuel use patterns, evaluation of SO_2:NO_x ratios, analyses of trajectories, and results of modeling.	*Moderate-High*: Continued research is necessary to quantify relationships between cause and effect.
Seasonal variations in observed acidity of precipitation.	*Moderate-High*: Seasonalities have been observed in acidity monitoring data. The seasonal trends in sulfate levels correlate well but nitrate levels do not seem to have an annual variation. Better definition of trends will shed light on source contributions.	*Low-Moderate*: Data currently being collected should be analyzed for seasonal as well as long-term trends.
Evidence of a trend toward increasing acidity of North American lakes.	*High*: Acidic lakes have been monitored in the Adirondacks of New York, in New England, and in Northern Minnesota. Evidence of trends is not well documented because of lack of historical data. One popular method of determining acidity increases is to note changes in lake productivity that may be a function of many other factors. The reliability of trend determinations is a function of the rate of change and the length of record.	*High*: Programs for the continuous monitoring of lake acidity over a broad geographic area are required.
Relative role of acid precipitation in contributing to acidity of lakes.	*High*: The factors contributing to lake acidification are not yet well understood. Local inputs from streams, bogs, and runoff over watershed areas may greatly enhance acid addition. Differing buffering capacities and effects of cumulative additions of acid by precipitation are important and differ widely between regions and even between lakes in the same region.	*High*: Local inputs of acids need to be quantified and the buffering process more fully investigated. Studies attempting this are currently underway.
Continuity of monitoring programs.	*High*: Agencies involved in these activities must be able to make firm commitments with support through programmatic processes.	*High*: Agencies involved in these activities must be able to make firm commitments with support through programmatic processes.

Effects.

How can we evaluate whether acidification of a lake has occurred? What effects can be predicted?	*High*: A quantification of the extent of the problem requires adoption of accepted criteria to identify an acidified water body or system.	*High*: A useful and generally accepted model needs to be developed. Such a model would require large amounts of background data on lake chemistry and other areas.
How fast will a given lake become acid? How and why do lakes vary in their susceptibility to acidification?	*High*: These data would enable a better quantification of the extent of acidification and provide a true measure of regions of special sensitivity.	*High*: A sensitivity analysis is required. Such an analysis requires a large data base. This analysis would have to include: the watershed/lake area; lake elevation; SO_4 reduction system; denitrification system; geological substrate; buffering of acid input by canopy, litter, soil, stream channels, etc.

(continued)

TABLE 6-1 (continued).

Item/Issue	Level of Uncertainty	Need for Further Research (Intensity Indicated Where Possible)
Long-term, comparative studies of carefully selected lakes. These studies should include quantitative analyses of vegetation, soils, geology, and hydrology of watersheds.	*Moderate*: Work done to date has concentrated on short-term acute impacts of acidification. Long-term effects of chronic acidic inputs have been relegated to a lower priority. Concern will undoubtedly increase to *High* as short-term impacts and mechanisms of acidification are resolved.	*High*: Long-term studies invariably are complex and costly.
Carefully planned laboratory and field studies need to be conducted to elucidate mechanisms and quantify effects of acidification on aquatic and terrestrial ecosystems. Specific areas of research could include:	*High*: Effects on aquatic species/ecosystems and terrestrial ecosystems are the primary concern regarding acidic precipitation. Both short-term acute effects and long-term chronic impacts are of importance.	*Variable*: Research effort will vary depending on the effects being investigated. Some questions could be resolved with a minimum expenditure of time and money; others would require extensive commitments of resources and personnel. Examples of areas of research and necessary levels of effort are listed below. Obviously there are many more areas that could be listed. These serve only to illustrate the complexity of the problems.
Bioassay of all forms of aquatic organisms for pH sensitivity and heavy metal toxicity.		Moderate-High
Physiological and behavioral studies of all kinds of organisms under stress at low pH via exposure to sulfuric and nitric acid in proportion to real world situation (e.g., 2:1, 1:1, 0.6:1 in various locations) and with exposure to neutralized salts.		Moderate-High
Physical and chemical characterization of dissolved organic matter.		Low-Moderate
Identification of chemical species of aluminum and toxicity of these species especially at pH 4-6.		Low-Moderate
Chemical transformations during draining through soils.		Moderate-High
The biomass and productivity of algal and invertebrate communities in acid waters.		Moderate-High
Mechanisms of the development of benthic sphagnum and algal mats and their influence on ionic exchange at the sediment/water interface.		Moderate-High

(continued)

TABLE 6-1 (continued).

Item/Issue	Level of Uncertainty	Need for Further Research (Intensity Indicated Where Possible)
Exchange of nutrients and metals in acid waters.		Moderate-High
Mercury transfer and accumulation processes in lakes.		Moderate-High
The components and characteristics of buffer systems in acid lakes.		Moderate-High
Response of fish to transient conditions.		Low-Moderate
Better evaluation of the sensitive stages in the life history of all kinds of organisms.		Moderate-High
Better elucidation of the mechanisms responsible for population extinction of both invertebrate fauna and fish.		High
Synergistic/antagonistic effects of pH, Ca, Al, and Mn.		High
The limits of acclimation and genetics in certain species.		Low-Moderate
Quantitative work on different acidity sources and sinks in the same soil needs to be conducted. For example:	High: As above. Especially of concern are effects of acidification of soils as this affects food crops.	Moderate-High: Again, level of effort depends on parameter to be quantified. However, elucidation of effects of acid precipitation on soil systems has in the past generally proved to require detailed effort.
H^+ contribution from plant acids and decomposition processes.		
Quantification of weathering and cation exchange processes and the effects of percolate composition.		
Rates of, and conditions for, nitrate and sulfate reduction.		
Mobilization of heavy metals in acid soils.		
Quantification of the effects of acidic rainfall on soil systems, in general, including:	High: As above.	Moderate-High: As above.
The theoretical soil chemistry of low level chronic acid inputs to soils.		
Applied measurements of acidification rates in laboratory and field studies.		

(continued)

TABLE 6-1 (continued).

Item/Issue	Level of Uncertainty	Need for Further Research (Intensity Indicated Where Possible)
Methods of evaluating the interaction of acid inputs and plant systems on soil acidity.		
Long-term effects of rainfall acidity on soil and soil/plant systems.	*High*: Concern is high about chronic impacts of acidic precipitation on terrestrial ecosystems, especially forest systems.	*Moderate-High*: Systems models could be generated to predict long-term effects. Actual in situ monitoring of long-term effects would be a major undertaking. Chronic effects could be simulated in the laboratory, but true extrapolation to the natural condition would be tenuous.
What are the effects on surface water quality of changing land use patterns?	*Moderate-High*: This area has not received enough study. Resolution of this question would contribute to a true quantification of the effects of acidic precipitation apart from other factors.	*Moderate-High*: A broad data base would be required.
What will be the ecological/biological consequences if acidic precipitation is eliminated?	*Low*: Concern at present is primarily directed at effects caused by the presence of acidic precipitation, not the converse. However, this area deserves careful study and evaluation, especially because some data indicate beneficial impacts of acidic precipitation.	*Moderate*: This area cannot be truly investigated without a clear picture of the impacts of acid rain on the systems affected.
What are the impacts of acidic precipitation on human health and well-being?	*High*: This question is of primary importance. Impacts should not only be assumed to represent health hazards but must also take into account loss of recreational and leisure activities, degradation of structures, monuments and other man-made objects, and effects on crops, forests, and aesthetic values, in general.	*Variable*: Minimal research efforts could quantify levels of trace toxic metals in potable waters and aquatic species and estimate impacts on health based on a broad background of toxicological information. Some estimates could easily be made to ascertain monetary loss resulting from crop damage, etc. The other areas of concern, many of which fall within the broad area generally recognized as quality of life, are less easily determined.
Mitigative Strategies		
How may the impacts of acid precipitation be mitigated?	*Moderate-High*: This question assumes all recognized impacts of acidic precipitation are detrimental. This has not been demonstrated to the satisfaction of all investigators involved in acidic precipitation research. Answering this question hinges on resolving most of the issues presented above, and involves evaluating both adverse and beneficial aspects of acidic precipitation.	*Moderate-High*: Mitigative strategies may be conceptualized with minimal difficulty once the effects of acid precipitation are quantified both chemically and physically.
The feasibility/desirability of liming as a mitigative strategy needs to be evaluated.	*Moderate*: Consideration of liming or other mitigative strategies has not received much attention in the U.S. to date.	*Variable*: This level of research efforts required to evaluate liming as a mitigative strategy depends on the aspect of the strategy under study. Benefits-cost analyses may be made with moderate effort assuming true impacts of acidification on a given lake or water body are known and if a realistic estimate of the logistics involved in liming a given area is available. Investigation into chemical, biological or ecological long-term effects of liming would be a major undertaking.

REFERENCES

1. Acid Rain Coordination Committee. The Federal Acid Rain Assessment Plan. Executive Office of the President, Council on Environmental Quality, Draft Report, August 1980.

2. Sigma Research Inc. In: Proceeedings of the Advisory Workshop to Identify Research Needs on the Formation of Acid Precipitation, Alta, Utah. Electric Power Research Institute, 1978.

Section 7

Current and Proposed Research on Acid Precipitation

Many organizations in the United States sponsor research on acid rain. According to one report,[1] several Federal agencies—including the Departments of Interior, Agriculture, and Commerce, as well as the Tennessee Valley Authority—will undertake research on acid precipitation and its effects during FY 1981. This section, however, will focus on the DOE's own research program and those of two other organizations: EPA, which sponsors the largest Federal research program on acid rain, and the Electric Power Research Institute (EPRI), whose research program has attracted international attention. Most of this section focuses on current acid precipitation research; it does not cover research being done by DOE, EPA, and EPRI on SO_x and NO_x control technology. Several large tables summarize each of the three organization's current acid rain research programs. The accompanying text highlights significant projects and indicates the direction in which each research program is headed.

DEPARTMENT OF ENERGY

The Department of Energy's current acid rain research program focuses on four areas: monitoring, atmospheric processes, ecological effects, and mitigative strategies. As Table 7-1 indicates, the Department estimates that it will allocate over $860,000 for acid precipitation research in FY 1980 and over $1.3 million in FY 1981. Over one-third of the funding in both years is devoted to monitoring work, whereas less than a fourth of the funding is allocated to ecological effects research.

Currently, DOE is participating in several large acid precipitation monitoring projects. One of these, the MAP3S program, was initiated by ERDA (now DOE) as a major sulfate pollution study. The original purpose of the study was to simulate the atmospheric effects of emissions from fossil-fueled power plants.

Between 1976 and 1978, funding from the study was used to establish a regional precipitation chemistry network in the eastern United States. This network, which is illustrated in Figure 7-1, was designed to provide the data needed to answer several important questions[2] including:

- What is the extent of the acid rain problem in the eastern United States?

- What are the relative contributions of natural versus anthropogenic pollutants to acid rain?

TABLE 7-1. DOE ACID RAIN RESEARCH PROGRAM; DOE OFFICE OF HEALTH AND ENVIRONMENTAL RESEARCH

Project title	Contractor	Funding ($1000s) estimates	
		FY 1980	FY 1981
Monitoring Research			
MAP3S Precipitation Chemistry Network	National Oceanic and Atmospheric Administration	89	89
Maintenance of a Rural Precipitation Chemistry Station at Whiteface Mountain	State University of New York at Albany	30	35
The Chemical Composition of Precipitation and Dry Deposition in the United States	Environmental Measurements Laboratory	295	320
Atmospheric Transport, Transformation, and Removal Process Research			
Atmospheric Pollution Scavenging	Illinois State Water Survey	250	250
A Climatological Analysis of Hubbard Brook Precipitation Chemistry Data	University of Toronto	90	0
Atmospheric-Terrestrial Environment Interaction of Energy Production Effluents	Oak Ridge National Laboratory	0	150
Ecological Effects Research			
Effects of Acid Precipitation on Forest Soils	Dartmouth College	66	58
Physiological Stresses of Acid Water on Fishes and Its Manifestation	Arizona University	41	41
Coal Combustion Effects in the Fort Union Basin	To be initiated	0	70
Air Pollution Effects on Food Quality	Pennsylvania State University	0	56
Mitigation Strategies Research			
Potential Acid Rain Mitigation Strategies	Argonne National Laboratory, ICF, Inc., Teknekron Research, Inc.	0	250
TOTAL		861	1,319

Source: Personal Communication between Lisa Baci, GCA, and Robert Beadle and Robert Kane, DOE, September 12, 1980, October 30, 1980.

- What are the relative contributions of wet and dry removal processes?

- What are the relative roles of emissions from power plants and other sources?

- Will proposed emission controls and energy use scenarios mitigate or increase the acid rain problem?

- Which ecosystems are most severely affected by acid rain, and which chemical constituents are responsible?

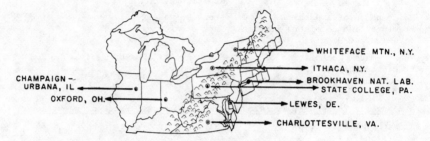

Figure 7-1. MAP3S precipitation chemistry network.

In 1978, the Office of Management and Budget directed DOE to transfer this project, as well as several others, to EPA. DOE continues to co-fund the MAP3S Precipitation Chemistry Network, however, and much of the work is still being conducted at DOE's national laboratories. The current direction of the project will be discussed in more detail in the subsection on EPA's research program.

DOE is also involved in another large monitoring program being conducted by the Environmental Measurements Laboratory. This project, which is funded at $295,000 in FY 1980 and $320,000 in FY 1981, focuses on the effects of changes in fuel-use patterns on the chemical composition of wet and dry deposition. Samples of wet, dry, and total deposition are collected monthly from seven sites located across the United States. Data from the project are published quarterly in the Environmental Measurements Laboratory Report.

As Table 7-1 indicates, DOE is also sponsoring several projects that examine the formation, effects, and mitigation of acid precipitation. DOE spent almost $350,000 for research on atmospheric transport, transformation, and removal processes in FY 1980 and estimates it will spend approximately $400,000 on this type of research in FY 1981. In FY 1980, DOE devoted over $100,000 to research on the ecological effects of acid rain, funding two projects--one that focused on the impacts of acid precipitation on forest soils and another that evaluated its impact on fish. In FY 1981, DOE plans to examine a wider range of ecological effects and expects to spend approximately $225,000 on this type of research. Finally, DOE has allocated $350,000 for work on acid rain mitigation strategies during FY 1981. Most of this money is being used to fund a project directed by Argonne National Laboratory, although $100,000 has been transferred to EPA for research on mitigation strategies.

DOE sponsors many research projects that may contribute to our understanding of acid rain even though they are not specifically designed to resolve uncertainties about acid precipitation. Table 7-2 outlines some of these indirect, or supporting, acid rain research projects for FY 1980 and FY 1981. As the table indicates, most of this research, funded at $897,000 for FY 1980 and $869,000 for FY 1981, focuses on the ecological effects of air pollutants,* particularly SO_2. DOE also sponsors research on SO_x and NO_x control technologies. It should also be pointed out that DOE is in the process of developing a data base regarding characterization of man-made sources.

In the future, DOE would like to focus its research program on establishing source-receptor relationships. Specifically, DOE would like to develop or expand its program in several areas including:*

- precipitation chemistry data analysis;

- diagnostic and predictive source-receptor analyses and modeling using regional/national emissions and precipitation data bases;

- atmospheric chemistry and physics of acid rain formation;

- acid precipitation effects on plants and soils with emphasis on trace element nutrients availability and losses in selected terrestrial, aquatic, and estuarine ecosystems;

- synergistic effects of dry and wet deposited acid species on vegetation and mechanisms of their interaction; and

- assessment and mitigation strategies.

ENVIRONMENTAL PROTECTION AGENCY

EPA's current acid rain research focuses on essentially the same areas as DOE's: environmental effects, monitoring, and atmospheric processes. In addition, as Table 7-3 indicates, EPA has also set aside some funding for two program support projects. According to EPA's recently compiled Inventory of Acid Rain Monitoring and Research Projects,[3] just under $4 million was devoted to acid rain monitoring and research in 1979; funding in 1980 is expected to exceed $5 million. In contrast to DOE, EPA spent over half of its budget for acid rain research in FY 1979 and FY 1980 on environmental effects and economies research and devoted only 10 to 15 percent of its funding to monitoring projects.

*Personal communication between Lisa Baci, GCA, and Robert Beadle, DOE, September 12, 1980.

TABLE 7-2. DOE--INDIRECT ACID RAIN RESEARCH

Project title	Contractor	Funding ($1000s) estimates	
		FY 1980	FY 1981
<u>Ecological Effects Research</u> (Including Supporting Acid Rain Research)			
Sorption of Pollution Gases by Soils	Iowa State	57	61
Air Pollution Effects on Food Quality	Pennsylvania State University	52	0
Likelihood Estimation of SO_2 Damage to Vegetative Alteration	Pennsylvania State University	77	83
Ecological Behavior and Effects of Energy Related Pollutants	Emory University	58	60
Transport, Fate, and Effects of Air Pollutants in Walker Branch Watershed	Oak Ridge National Laboratory	320	345
Air/Earth Cycling of Pollutants	Savannah River Laboratory	235	200
Effects of SO_2 on Light Modulation of Plant Metabolism	University of Illinois	35	50
Biochemical Effects of SO_2 on Plants	Michigan State University	63	70

TABLE 7-3. EPA ACID RAIN RESEARCH PROGRAM[3]

Project title	Affiliation of principal investigator(s)	Funding ($1000s) estimates	
		FY 1979	FY 1980
Environmental Effects and Economics Research			
Research on the Effect of Acid Precipitation on Aquatic and Terrestrial Ecosystems	North Carolina State University	500	648
Short-Term Research Program on Environmental Effects of Acid Rain	Brookhaven, Purdue, Univ. of Calif.-Riverside, Penn State, Oregon State, MITRE Corporation	115	0
Effects of Acid Precipitation on Crops and Forests	EPA Environmental Research Laboratory, Corvallis, OR, Oregon State University	270	290.2
Greenhouse Microcosm Studies on Effects of Sulfuric and Nitric Acid Rain on Selected Agricultural Crops	EPA Environmental Research Laboratory, Corvallis, OR	0	87.3
Effects of Acid Precipitation on Soil Biological Processes	EPA Environmental Research Laboratory, Corvallis, OR	0	70.5
Effects of Acid Rain on Mycorrhizal Fungi and Growth of Mycorrhizal vs. Non-Mycorrhizal Conifers	EPA Environmental Research Laboratory, Corvallis, OR	0	10
Assessment of the Sensitivity (Susceptibility) Index Concept for Evaluating Resources at Risk from Atmospheric Pollutant Deposition	The Institute of Ecology, Butler University	0	50
Characterization and Quantification of the Transfer, Fate and Effects of SO_x, NO_x and Acid Precipitation on Forest Ecosystems Representative of the Tennessee Valley Region	Tennessee Valley Authority, Oak Ridge National Laboratory	250	225
Effects of Acid Rain on Terrestrial Ecosystems	Brookhaven National Laboratory	258	258
Acid Precipitation Effects on Crops	Argonne National Laboratory	10	20
Effects of Acidified Precipitation on Fish Resources	U.S. Fish and Wildlife Services	0	128

(continued)

TABLE 7-3 (continued)

Project title	Affiliation of principal investigator(s)	Funding ($1000s) estimates	
		FY 1979	FY 1980
Ecological Effects of Coal Combustion: Interactive Effects of Vegetation to SO_2, Ozone and Acid Precipitation	Oak Ridge National Laboratory	131	131
The Mobilization and Transportation of Soil and Sediment Components into Pollutants by "Acid Precipitation" and Related Factors	EPA Environmental Research Laboratory, Duluth, MN	93	182[a]
Impacts of Airborne Pollutants on Wilderness and Park Areas of Northern Minnesota, Wisconsin and Michigan	EPA Environmental Research Laboratory, Duluth, MN	150	120
Generalization of Water Quality Criteria Using Chemical Models	University of Minnesota	10[a]	11.1
Experimental Field Studies to Evaluate the Effects of Acidification on a Stream Ecosystem	Cornell University	0	59
Coal Fired Steam Plants: Human and Environmental Exposure to Air and Water Pollution	University of Wisconsin	57.6[a]	162.4
Effects of Precipitation and Solutes on the Surface and Groundwater Quality in the Filson Creek Watershed Area	USGS	7	5
Acid Effects of Fathead Minnows and Macroinvertebrates in Outdoor Experimental Channels	EPA Monticello Ecological Research Station	36.5[a]	0
Effect of Acid on Zooplankton Community Structure and Functional Response in an Experimental Mesocosm and Northern Minnesota and Wisconsin Lakes	University of Minnesota	5	10
Explorations into the Effects of Acid Conditions on Aquatic Invertebrates	University of Minnesota	5.5	0

(continued)

TABLE 7-3 (continued)

Project title	Affiliation of principal investigator(s)	Funding ($1000s) estimates	
		FY 1979	FY 1980
The Environmental Impact of Energy-Related Organic Compounds on Aquatic Life	University of Minnesota	21.7[a]	60
Examination of Fish Recruitment in 130 Wisconsin Lakes (pH 4.5-7.5)	EPA, Environmental Research Laboratory, Duluth, MN	0	35
Impact of Acid Rain on Drinking Water in New England and New York	To be Selected	0	100
Budget of Man-made Sulfur, Nitrogen and Hydrogen Ions over the Eastern United States and Southeastern Canada Airshed	Washington University	15[a]	20[a]
Design of Economic Methodology for Assessing the Benefits of Controlling Acid Precipitation	University of Wyoming	52	52
Monitoring and Quality Assurance Research			
Monitoring System Support to Acid Rain Program[b]	EPA Environmental Research Monitoring Systems Laboratory	130	235
Quality Assurance Guidelines and Reference Samples	Rockwell EMSC	0	65
Monitoring System Support to Acid Rain Program[b]	EPA Environmental Monitoring Systems Laboratory	0	150
Precipitation Chemistry Network	Battelle Pacific Northwest Laboratories	160	160
Precipitation Chemistry in Western Oregon	EPA Environmental Research Laboratory, Corvallis, OR	0	32.2
Acid Rain: National and International Assessment of Potential Impacts	NOAA/ERL/ARL	0	50
Dry Deposition Studies	Argonne National Laboratory	105	95

(continued)

TABLE 7-3 (continued)

Project title	Affiliation of principal investigator(s)	Funding ($1000s) estimates	
		FY 1979	FY 1980
Test Dry Deposition Monitors and Techniques for Aerosols and Gases	EPA, Environmental Sciences Research Laboratory	0	25
Operation of Dichotomous Sampler at MAP3S Site	University of Virginia	10	4
Atmospheric Processes Research			
MAP3S/RAINE PROGRAM			
• MAP3S Program Direction	Battelle PNL	39[a]	78[a]
• MAP3S Modeling Studies	Battelle PNL	63[a]	126[a]
• Precipitation Scavenging	Battelle PNL	340	340
• Laboratory Research in MAP3S	Battelle PNL	63[a]	126[a]
• Chemical Characterization of Aerosols	Argonne Nat'l Lab.	10[a]	25[a]
• Boundary Layer Investigations	Argonne Nat'l Lab.	40[a]	75[a]
• Transport and Transformation	Brookhaven Nat'l Lab.	130[a]	260[a]
• MAP3S Central Data Coordination	Brookhaven Nat'l Lab.	50[a]	100[a]
• Aircraft Operations	Brookhaven Nat'l Lab.	50[a]	100[a]
• Modeling and Analysis	Brookhaven Nat'l Lab.	55[a]	110[a]
Precipitation Scavenging of Pollutants	UCLA, Atmospheric Sciences Dept.	78	85
Refinement, Verification, and Application of a Long-Range Transport Model of SO_2 and Sulfate	Colorado State University	60	75
Precipitation Chemistry Field Program	U.S. Military Academy	25	25
Adaptation of EURMAP Model for Eastern United States	SRI, International	69	50
Program Support Projects			
Acid Rain Program Support	The MITRE Corporation	45	94
Acid Rain Program Support	Teknekron Research, Inc.	55	70

[a] Portion of total project funding devoted to acid rain research.

[b] Although these projects have identical titles, brief abstracts contained in EPA's Acid Rain Research Inventory indicate they have somewhat different objectives.

EPA's research on environmental effects and economics examines a wide range of topics. Projects currently underway investigate the impact of acid rain on agriculture, forests and other terrestrial ecosystems, aquatic ecosystems, drinking and ground water supplies, and materials' deterioration. In FY 1980, the agency is spending approximately $2.7 million analyzing the environmental consequences of acid precipitation. A few of the key research projects in this area are summarized briefly below.

Effect of Acid Precipitation on Aquatic and Terrestrial Ecosystems

North Carolina State University is managing this multipart program that seeks to:

- determine the geographic distribution of sensitive aquatic and terrestrial ecosystems and those exhibiting symptoms of damage, and identify the current extent of damage;

- determine actual and potential effects on terrestrial and aquatic components of lake-watershed ecosystems, and develop models linking ecological response to acid precipitation inputs; and

- determine effects on native and commercial vegetation.

A variety of research proposals have been funded through this project, including studies on the effects of acidification on softwater lakes in Florida and the effects of changing patterns of acidic precipitation on the quality and yield of major agricultural crops of northeastern United States. One study funded under this project examines the relationship between acid rain and materials damage on stone and is part of an international effort to study adverse environmental effects of acid deposition on historic and artistic stone monuments.[4]

Effects of Acid Precipitation on Crops and Forests

This project, which is being undertaken at EPA's Environmental Research Laboratory in Corvallis, Oregon, has two objectives: (1) determine effects of sulfuric and nitric acid rain on foliage and yield of important farm crops, and (2) estimate effects of sulfuric acid rain on forest productivity and study adverse environmental effects of acid deposition on historic and artistic stone monuments.[4]

Characterization and Quantification of the Transfer, Fate, and Effects of SO_x, NO_x and Acid Precipitation on Forest Ecosystems Representative of the Tennessee Valley Region

The purpose of this project, which is being jointly conducted by the Tennessee Valley Authority and the Oak Ridge National Laboratory, is to compare, characterize, and quantify the transfer, fate, and effects of sulfur and nitrogen compounds entering two forest watersheds on the Cumberland Plateau.

A report (EPA-600/7-79-053) describing the objectives, facilities, and ecological characteristics of the study sites has been completed. Research emphasis is currently shifting from baseline quantification to deposition modeling and comparative mass balance development. Two models—one describing SO_x and NO_x deposition on forest canopies and another examining ecosystem response to changing atmospheric inputs—are scheduled to be produced before the project is completed in late 1984.

Effects of Acid Rain on Terrestrial Ecosystems

This project, conducted by the Brookhaven National Laboratory, is examining the effects of acid precipitation and acid aerosols on (1) forest and crop plants of northeastern United States, (2) microbial decomposition processes, (3) the nitrogen cycle, and (4) Rhizobium-legume symbiosis. The project is expected to produce integrated models of (1) forest nutrient cycling and growth and (2) materials routing during plant growth and is scheduled to continue for several more years.

As Table 7-3 indicates, EPA is involved in a number of monitoring projects. Currently, the agency is solely or partially supporting several monitoring networks including the MAP3S Precipitation Chemistry Network, which is outlined in Figure 7-1, the National Atmospheric Deposition Project Network (NADP), which will eventually include more than 50 monitoring stations nationwide, and a 15-station global network established by the World Meteorological Organization (WMO). According to EPA's Inventory of Acid Rain Monitoring and Research Projects,[3] the agency spent $405,000 on monitoring programs in FY 1979 and expects to more than double this figure in FY 1980.

Through one of its monitoring system support programs, EPA has established a central data bank that stores data from a variety of U.S. monitoring networks, including those of MAP3S, NADP, TVA and EPRI, as well as data from the WMO and Canadian monitoring networks. This data bank, which is located at EPA's Environmental Monitoring Systems Laboratory in Research Triangle Park, NC, became operational in December 1979. Future goals of the monitoring system support projects include: (1) establishing a core monitoring network for use in identifying major variables and developing a reliable acid rain monitoring system and (2) developing and implementing a program to ensure data quality.

Most of EPA's research on atmospheric processes is being conducted through the MAP3S/RAINE program that the Office of Management and Budget transferred to EPA from the Department of Energy. In FY 1980, slightly less than half of the MAP3S budget of over $2.5 million was spent on acid rain related research. In FY 1981, EPA plans to allocate all of the MAP3S budget of approximately $3 million to acid rain research.* The MAP3S research has many diverse objectives ranging from the development of a series of models capable of investigating problems with long-range pollution transport and acid species deposition to the development of advanced pollution monitoring equipment capable of airborne applications.

---

*Personal communication between Lisa Baci, GCA, and David Bennett, EPA, September 18, 1980.

More information on the projects listed in Table 7-3 is available in the <u>Inventory of Acid Rain Monitoring and Research Projects</u>, just compiled by EPA's Office of Research and Development.³ This inventory contains a one- to two-page summary of each project including information on funding, research objectives and approach, and expected output and delivery dates. EPA plans to update this inventory annually unless the Federal Acid Rain Coordination Committee follows through with its proposal to establish a national, computerized acid rain research inventory.*

THE ELECTRIC POWER RESEARCH INSTITUTE

The Electric Power Research Institute (EPRI) is a nonprofit research arm of the U.S. electric utility industry. Over the past 3 years, EPRI has spent over $5 million on acid rain research and anticipates spending another $10-15 million over the next 5 years.⁵ As Table 7-4 indicates, EPRI's ongoing research concentrates on both the ecological effects and the environmental physics and chemistry of acid deposition. All of EPRI's acid rain research is being performed under contract, much of it with universities and DOE's national laboratories.

Several of EPRI's research projects have attracted considerable attention, particularly the Integrated Lake-Watershed Acidification Study and the study on the Fate of Atmospheric Emission Plume Trajectory over the North Sea. Both of these studies are summarized briefly below.

In the Integrated Lake-Watershed Acidification Study, scientists from several distinguished universities and research organizations are studying three lakes of differing acidities located in the Adirondack region of New York. One of the major goals of this multimillion dollar project is to produce a model that predicts how acid rain interacts with elements of the environment to affect lake acidity. The project, which involves an extensive program of field measurements, is scheduled to be completed in 1983. If the model (a preliminary version of which was recently tested in Sweden) proves satisfactory, EPRI hopes to conduct regional field studies using the model in other areas of the United States.†

The study on the Fate of Atmospheric Emission Plume Trajectory over the North Sea is a two-phase project being carried out by the Central Electricity Research Laboratory of Great Britain. The first phase of the project involved the development of two instruments: (1) a continuous tracer analyzer able to operate when airborne and (2) a cloud water collector that does not collect interstitial matter. Both these instruments are currently in use in the second phase of the project, which involves tracing airborne sulfur and nitrogen emissions from a coal-fired power plant as they move across the North Sea towards Norway. According to Dr. Ralph Perhac who is managing the project, EPRI hopes to "determine, for the first time, exactly how acidity is formed in the raindrops and ascertain what role, if any, power plants play in the

---

*Personal communication between Lisa Baci, GCA, and David Bennett, EPA, September 18, 1980.

†Personal communication between Lisa Baci, GCA, and Myra Fraser, EPRI, September 10, 1980.

TABLE 7-4. EPRI ACID RAIN RESEARCH PROGRAM

Project number and title		Contractor(s)
Ecological Effects Research		
RP1109	Integrated Lake-Watershed Acidification Study	RPI, Cornell, Virginia, Dartmouth, Colgate, Smith, USGS, Brookhaven, Tetra Tech.
RP1313	Photosynthetic Response to Gaseous Pollutants	Stanford University
RP1632	Microcosm Evaluation of Acidic Deposition on Forest Ecosystems	Tennessee Valley Authority
RP1635	Sulfur Effects on Grasslands, A System Analysis	Colorado State University
RP1812	Effects of Acid Precipitation on Agricultural Crops - Northeast	Boyce-Thompson Institute
RP1908	Effects of Acid Precipitation on Agricultural Crops - Midwest and Southeast	Argonne National Laboratory Oak Ridge National Laboratory
RP1813	Effects of Acid Rain on the Nutrient Status of Forest Ecosystem	Oak Ridge National Laboratory
RP1907	Forest Canopy Interactions with Acid Rain	Oak Ridge National Laboratory
Environmental Physics and Chemistry Research		
RP1155-1	Acidic Precipitation in the Adirondack Region	Rensselaer Polytechnic Institute
RP1630-2	Eastern Regional Air Quality Studies - Precipitation Chemistry Measurements	Rockwell International Inc.
RP1311	The Fate of Atmospheric Emission Plume Trajectory over the North Sea	Central Electricity Research Laboratory of Great Britain
RP1434	Effects of Aerosols and Cloud Droplets on Nighttime Transformation of Sulfur Oxides	Desert Research Institute

Source: Personal communication between Lisa Baci, GCA, and Myra Fraser and John Jansen, EPRI, September 10 and 15, 1980.

process."[6] Field work for the $965,000 project is scheduled to be completed in December 1980.

EPRI estimates that it will spend over $13 million between 1980 and 1985 on ecological research alone.[7] Most of this money is divided fairly evenly between studies in three areas--lake acidification, crop yield and quality, and forest yield and quality--with each area allocated between $3.2 and $3.6 million. EPRI estimates it will spend between $1.5 and 1.6 million for studies in two other areas of ecological effects research: grassland yield and quality and aquatic biota. In addition to the projects listed in Table 7-4, all of which are approved or actually underway, EPRI hopes to fund studies in the following areas:*

- evaluation of nitrogen deposition on forests,
- evaluation of SO_2 deposition on forests,
- effects of acid rain on aquatic biota, and
- effects of acid deposition on agricultural soils.

Although most of EPRI's current projects in environmental physics and chemistry are scheduled to end shortly, EPRI does plan to fund new projects in this area in the future. EPRI is currently making final project selections for the upcoming year. In the future, it hopes to investigate better regional transport models, undertake more detailed precipitation chemistry analyses, and conduct more field measurements and mass budget analyses of cloud chemistry in the United States.†

---

*Personal communication between Lisa Baci, GCA, and Myra Fraser, EPRI, September 10, 1980.

†Personal communication between Lisa Baci, GCA, and John Jansen, EPRI, September 15, 1980.

REFERENCES

1. Acid Rain Coordination Committee. The Federal Acid Rain Assessment Plan. Draft, 1980.

2. Department of Energy. FACT SHEET, The Acid Precipitation Problem. Draft.

3. Office of Research and Development. Inventory of Acid Rain Monitoring and Research Projects Spanning FY 1979 to FY 1980. U.S. Environmental Protection Agency. Draft, 1980.

4. Office of Research and Development. Research Summary, Acid Rain. U.S. Environmental Protection Agency, 1979.

5. Perhac, R. M. Testimony for the Electric Power Research Institute before the Subcommittee on Environmental Pollution of the Senate Committee on Environment and Public Works. March 19, 1980.

6. EPRI Research Project to Study Acid Rain. Journal of the Air Pollution Control Association, December 1979.

7. Electric Power Research Institute. Ecological Effects Program, Acid Deposition Research. 1980.